Technology Roadmapping for Strategy and Innovation

Martin G. Moehrle, Ralf Isenmann,
and Robert Phaal (Eds.)

Technology Roadmapping for Strategy and Innovation

Charting the Route to Success

Springer

Editors
Prof. Dr. Martin G. Moehrle
IPMI - Institut für
Projektmanagement und Innovation
Universität Bremen
Bremen
Germany

Prof. Dr. Ralf Isenmann
Fraunhofer Institute for Systems
and Innovation Research ISI
Competence Center Innovation
and Technology Management and Foresight
Karlsruhe
Germany

Dr. Robert Phaal
Centre for Technology Management
Institute for Manufacturing
Department of Engineering
University of Cambridge
Cambridge
United Kingdom

ISBN 978-3-642-33922-6   e-ISBN 978-3-642-33923-3
DOI 10.1007/978-3-642-33923-3
Springer Heidelberg New York Dordrecht London

Library of Congress Control Number: 2012948328

© Springer-Verlag Berlin Heidelberg 2013
This work is subject to copyright. All rights are reserved by the Publisher, whether the whole or part of the material is concerned, specifically the rights of translation, reprinting, reuse of illustrations, recitation, broadcasting, reproduction on microfilms or in any other physical way, and transmission or information storage and retrieval, electronic adaptation, computer software, or by similar or dissimilar methodology now known or hereafter developed. Exempted from this legal reservation are brief excerpts in connection with reviews or scholarly analysis or material supplied specifically for the purpose of being entered and executed on a computer system, for exclusive use by the purchaser of the work. Duplication of this publication or parts thereof is permitted only under the provisions of the Copyright Law of the Publisher's location, in its current version, and permission for use must always be obtained from Springer. Permissions for use may be obtained through RightsLink at the Copyright Clearance Center. Violations are liable to prosecution under the respective Copyright Law.
The use of general descriptive names, registered names, trademarks, service marks, etc. in this publication does not imply, even in the absence of a specific statement, that such names are exempt from the relevant protective laws and regulations and therefore free for general use.
While the advice and information in this book are believed to be true and accurate at the date of publication, neither the authors nor the editors nor the publisher can accept any legal responsibility for any errors or omissions that may be made. The publisher makes no warranty, express or implied, with respect to the material contained herein.

Printed on acid-free paper

Springer is part of Springer Science+Business Media (www.springer.com)

Acknowledgements

The editors are grateful to the many people who have contributed to this book. First of all, we would like to mention the authors who allowed us to provide a broad spectrum of papers covering all relevant aspects of technology roadmapping. They all were extremely constructive and had the patience to give us the time for a careful development of the book. Dipl.-Geogr. Christian Hanke provided significant help regarding various editorial aspects. He carefully examined the papers, unified organizers, layout and references, and was a reliable contact partner for the involved authors and editors. Mag. Kirsten Kueven, who is not only a patient discussion partner but also a passionate friend of the English language, did a lot of linguistic polishing. Sarah Oelund and Umar Aziz, both students of a master course in Engineering and Management, helped us to create a coherent graphical design for all papers, linking them not only in terms of content but also by visual appearance. Thomas Lehnert and Dr. Jan-Philip Schmidt acted as our editors at Springer Science; he encouraged us to realize our concept for a book that we see as a comprehensive compendium for technology roadmapping. Last not least, we wish to thank our partners in industry and other organizations who shared their experience on technology roadmapping with us and helped us to select a portfolio of ambitious and, at the same time, practice-oriented papers.

Bremen, Karlsruhe and Cambridge Martin G. Moehrle
September 2012 Ralf Isenmann
Robert Phaal

Contents

Basics of Technology Roadmapping 1
Martin G. Moehrle, Ralf Isenmann, Robert Phaal

Part 1: Institutional Reference for Technology Roadmapping

Technology Management and Roadmapping at the Firm Level 13
Robert Phaal, Clare Farrukh, David R. Probert

Networked Innovation: Using Roadmapping to Facilitate Coordination, Collaboration and Cooperation 31
Irene J. Petrick

Technology Roadmapping on the Industry Level: Experiences from Canada 47
Geoff Nimmo

Roadmapping as a Responsive Mode to Government Policy: A Goal-Orientated Approach to Realising a Vision 67
Clive I.V. Kerr, Robert Phaal, David R. Probert

Part 2: Processes of Technology Roadmapping

Fast-Start Roadmapping Workshop Approaches 91
Robert Phaal, Clare Farrukh, David R. Probert

Technological Overall Concepts for Future-Oriented Roadmapping 107
Günther Schuh, Hedi Wemhöner, Simon Orilski

Scenario-Based Exploratory Technology Roadmaps – A Method for the Exploration of Technical Trends 123
Horst Geschka, Heiko Hahnenwald

TRIZ-Based Technology Roadmapping 137
Martin G. Moehrle

Development of Technology Foresight: Integration of Technology Roadmapping and the Delphi Method 151
Daisuke Kanama

The Innovation Support Technology (IST) Approach: Integrating Business Modeling and Roadmapping Methods 173
Hitoshi Abe

Part 3: Implementing Technology Roadmapping

Implementing Technology Roadmapping in an Organization 191
Nathasit Gerdsri

Innovation Business Plan at Siemens: Portfolio-Based Roadmapping to Focus on Promising Innovation Projects Right from the Beginning ... 211
Babak Farrokhzad, Claus Kern, Meike de Vries

Exploratory Roadmapping: Capturing, Structuring and Presenting Innovation Insights 225
David A. Beeton, Robert Phaal, David R. Probert

Part 4: Linking Technology Roadmapping to Other Instruments of Strategic Planning

Long-Term Innovation Management – The Balanced Innovation Card in Interplay with Roadmapping 243
Rainer Vinkemeier

Strategic Visioning – Future of Business 257
Volkmar Doericht

Linking Technology Roadmapping to Patent Analysis 267
Sungjoo Lee

Author Index 285

Basics of Technology Roadmapping

Martin G. Moehrle, Ralf Isenmann, and Robert Phaal

Corporate technology managers are faced with a wide range of responsibilities: Apart from being in charge of the acquisition, preservation, protection and application of technological competencies, they are expected to attend to a preferably solid and market-oriented technological positioning of their company (for the scope of technology management see for instance Burgelman, Christensen and Wheelwright 2004). This accumulation of tasks has given rise to a need for the projection of a technology's temporal development, including its prevalently heterogeneous connections as well as the derivation of activities which serve to support or even improve a company's technological standing. Technology roadmapping represents an ideal method of dealing with the latter two of the abovementioned concerns in an integrative way.

- *The concept of technology roadmapping obviously takes its name from the metaphorical image of a roadmap. In effect, the business is seen as a kind of vehicle travelling through partly known, partly unknown territory, driven by somebody who might require navigational assistance.*
- *Technology roadmapping is applied in many different forms, ranging from purely internal and in most cases highly confidential revisions to the competitively strategic publication of customer information.*

Martin G. Moehrle
IPMI - Institut für Projektmanagement und Innovation, Universität Bremen,
Wilhelm-Herbst-Str. 12, 28359 Bremen, Germany
e-mail: martin.moehrle@innovation.uni-bremen.de

Ralf Isenmann
Fraunhofer Institute for Systems and Innovation Research ISI,
Competence Center Innovation and Technology Management and Foresight,
Breslauer Straße 48, 76139 Karlsruhe, Germany
e-mail: ralf.isenmann@isi.fraunhofer.de

Robert Phaal
Centre for Technology Management, Institute for Manufacturing,
Department of Engineering, University of Cambridge, 17 Charles Babbage Road,
Cambridge, CB3 0FS, United Kingdom
e-mail: rp108@cam.ac.uk

1 Framework, Processes, Implementation and Links: Four Parts of the Book

This book contains a description of technology roadmapping in four major parts, providing expert knowledge on (i) framing/embedding of technology roadmapping, (ii) processes of technology roadmapping, (iii) implementing technology roadmapping and (iv) linking technology roadmapping to other instruments of strategic planning (see Figure 1).

- In part 1 of the book the *institutional reference* for technology roadmapping will be explored. We will show that technology roadmapping can be used in different contexts, at different company and industry oriented levels, and within different management frames.
- In part 2 we will introduce several *processes* that have been established for successful technology roadmapping. To frame the processes we will use a starting point matrix of market driven versus technology driven and explorative versus directed approaches.
- In part 3 promising ways are described how technology roadmapping could be *implemented* and institutionalized.
- In part 4 we will *link* technology roadmapping to other planning methods. Especially interaction with corresponding methods along strategic planning and innovation planning are of specific interest.

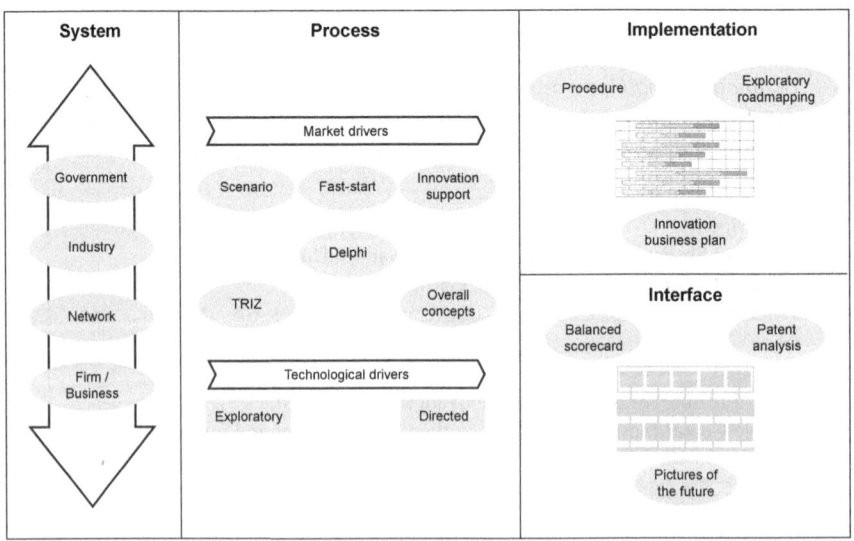

Fig. 1 Framework of the book

2 A Work of Art Inspires an Analogy

If you have ever covered any greater distance by car, it is quite likely that this involved the use of a very special work of art: a roadmap. Even the most commonplace edition shows:

- the localities that are to be found in a certain region,
- in how far these localities are connected by traffic routes,
- the size and inner structure of localities,
- the distance between a locality and a defined point of reference,
- which localities are close to and which remote from one another,
- whether there is a direct route of access between two localities or an unavoidable detour,
- the capacity of pathways between individual localities (according to road classification),
- whether there are any obstacles to overcome in getting from one locality to the other (e.g. by ferry or rail-crossing),
- interchanges between roads and other means of transport (airports, railway stations, harbours), and finally,
- characteristics of the surrounding landscape (lakes and rivers, elevations, vegetation).

Perhaps it is this practically tangible diversity of functions that motivates executives and experts alike to devise roadmaps for their particular fields of technology. After all, the similarities between the features of a roadmap and the requirements of medium-range control in technology management happen to be striking. And, of course, a technology manager needs to know:

- what technologies (and thus options concerning new products and services as well as new versions of established products or services) are available in a certain area,
- how strong the connections between existing technologies are,
- how powerful these technologies happen to be and what their inner structure is like,
- the distance (in terms of time and technical state) between a technology and a defined point of reference (e.g. a company's actual technological level),
- which technologies are proximate and which remote from one another,
- whether there are any direct links between individual technologies or detours that have to be taken into account,
- the capacity of connections between individual technologies,
- what problems have to be overcome on the way to a certain technology,
- whether there are any convergences with other meta-technologies, and
- what characterizes the surrounding technological territory.

Accordingly, a technology roadmap is nothing less than a graphical representation of technologies, often relating objects like products or competencies and the connections that have evolved between them in the course of time. The activities required in generating and updating this kind of representation are referred to as technology roadmapping (see Table 1 for related terms). Both topics, i.e. the method of technology roadmapping and the qualities of its output, will be described in this publication.

Table 1 Glossary – related terms (see porter and cunningham 2005, p. 18)

Technology monitoring and technology watch ...	... refer to operational activities that chiefly aim at identifying significant cross-business and often internationally oriented developments of the technological kind
Technology prognosis, technology foresight and technology projection ...	... are closely related to technology roadmapping. Especially long-range forecasts, like those contained in the Japanese Delphi-Report, provide a substantial orientation-aid to executives and experts concerned with technology roadmapping.
Technology efficiency analysis and technology impact assessment ...	... extend beyond the business level to general economic aspects as well as social and individual phenomena connected with the implementation of new technologies. Corresponding surveys function as an important corrective in the context of operational considerations.

3 'Technology Roadmapping' or 'Roadmapping'?

The concept of technology roadmapping, which is central to this book, can be specified in two different ways:

- On the one hand, there is a rather stringent definition including only roadmapping activities that focus on product- or process-related *technologies*. According to this interpretation of the term, product roadmaps, project roadmaps and function roadmaps are on equal footing with technology roadmaps.
- On the other hand, the concept can be interpreted in a less rigorous manner, encompassing all roadmapping activities concerned with technologies, products, processes, functions, market agents, competencies, projects and further aspects. This perspective can be traced back to the multi-level model of technology roadmapping as presented by the European Industrial Research Management Association (1998). Some authors would rather refer to this comprehensive concept by means of the more neutral term 'roadmapping' (without the technology prefix).

In this book technology roadmapping is primarily referred to in its broader sense. Nevertheless, the editors have accepted the stricter specification along with the use of the expression roadmapping as a generic term where authors mean to sketch out differences between roadmap types.

4 Variants of Technology Roadmapping

Several variants of technology roadmapping find application in operational activities today. The exact choice depends on what *reference objects* are involved as well as on the actual *goal* which can mostly be associated with certain interest groups or alliances of interest groups.

Technology roadmapping may involve various reference objects. A threefold classification thus appears to be recommendable: (i) roadmapping for central pace-setting and key technologies, (ii) roadmapping for application systems, and (iii) roadmapping for a company's or an industry's general productive output.

Some technologies exert quite a compelling pull concerning their application. Of late, this has been particularly noticeable in fuel cell technology, genetic technologies and internet technology. Therefore, it makes sense to create roadmaps for such pace-setting key technologies, from which a company's management is able to draw conclusions about the potential of different application systems. For instance, progress in fuel cell technology can have a revolutionary impact on the automotive industry as well as on minor or major electricity suppliers.

Apart from pace-setters and key technologies, technology roadmapping may also focus on application systems. Subjects like tomorrow's office, tomorrow's vehicle, tomorrow's building or even tomorrow's internal professional training can just as well be dealt with by means of a roadmap. The fact that this kind of subject would naturally involve a variety of individual technologies accounts for the differentiation from the "basic" roadmapping definition.

A company's or an industry's productive output represents a further possible focus of technology roadmapping. In this context the term productive output refers to the established and future range of products combined with services. Again, this may concern various technologies, so there is no restriction to one particular field of application; and in most cases this kind of technology roadmap would be complemented by a respective product roadmap. In spite of appearing less evident, the latter can also apply to type 1 and 2 roadmaps.

5 Purposes of Technology Roadmapping

These different reference objects of technology roadmapping account for much of its diversity, part of which can furthermore be attributed to interest group-related issues. After all, goals and purposes may vary, depending on which and how many interest groups are involved in the technology roadmapping process.

Some technology roadmaps serve only one purpose: the supervision of intra-corporate R&D units. In such cases it is necessary to differentiate between roadmaps which are drawn up by the unit itself to formulate a kind of self-specified target, and those compiled on behalf of the management by internal or external consultants for the revision of the corporate R&D unit's form and structure. Generally, both categories are treated as strictly confidential.

Technology roadmaps can also be used to facilitate the co-ordination of different staff functions. For instance, the wide-spread distrust between R&D and marketing units may be mitigated by means of a conjoint roadmap to which R&D contributes technological factors while marketing staff bring in a product-related perspective.

Another field in which roadmaps prove to be helpful is that of competitive strategy. In this case, the interest groups involved are the company's marketing unit and its customers. The strategic utilization of roadmaps is primarily reflected by the company's announcement policy. A striking example of this can be observed in the computer industry, where the Microsoft Corporation regularly succeeds in deterring consumers from purchasing a competitive product by specifically announcing the upcoming launch of a similar article (so-called 'vaporware').

Furthermore, technology roadmaps enable the co-ordination of intra- and extra-corporate R&D activities. This function especially presents itself where extensive co-operations or a high level of external procurement are immanent. Eventually, individual companies have the option to join forces in devising a technology roadmap that supports their common orientation. This has happened very prominently in the semiconductor industry. Here, it is the international consortium SEMATECH that issues technology roadmaps for the entire sector (see Sematech, 2011).

6 Limitations of Use

Technology roadmaps are largely concerned with prognosticating technical developments and their interactions. On the whole, forecasts tend to be rather uncertain, and the dictum that "planning replaces accident by error" is not altogether unjustified. Hence, it makes sense to inquire what *limitations* of use apply to technology roadmapping.

There can be no doubt that it is virtually impossible to prognosticate a fundamental technological break-through or findings of seminal importance with reliable exactitude. The discovery of x-rays, the exploration of genetics, or (to state a more recent example) the detection of fullerenes (i.e. carbon molecules consisting of 60 atoms each, which adopt a football-like shape) – who would have dared to predict all that with any claim to precision? As concerns this elementary question, even the Delphi Reports have no answers to offer.

However, fairly reliable prognoses can be attained in the wake of such discoveries, regarding their further development. This is the stage at which technology managers make use of technology roadmapping. For instance, the documented existence of fullerenes could now be followed by the search for options of advancement and commercialization that might serve as the theoretical framework of a technology roadmap.

7 New Aspects of Technology Roadmapping

At the close of this introductory chapter it should be mentioned that the technique of roadmapping is not entirely a novelty. Practised corporate executives and experts have been thinking in roadmaps for a long time without reverting to any graphical representation thereof. Nevertheless, the current discussion of roadmapping is marked by several new aspects:

- The "extraction from the mind", i.e. the physical documentation of technology roadmapping combined with a communicative purpose,
- consequently, the generation of roadmaps across and beyond the boundaries of departments or, indeed, companies as well as other institutions, and
- the use of intelligible tools such as scenario planning in technology roadmapping.

These three aspects can practically be regarded as characteristic features of technology roadmapping. While the first feature listed above amounts to an *explication of the implicit* by way of organized documentation with specific communicative purposes, the second feature is concerned with the options of *institutionalisation*, i.e. the organisation of technology roadmapping, ranging from a "closed shop procedure" inside the corporate R&D unit to a more open and overt approach that includes the participation of suppliers, co-op partners and key clients as external stakeholders. The third feature mentioned here underlines the *instrumental linking function* that technology roadmaps are able to fulfil. Increasingly often roadmaps are methodically combined with other tools such as balanced scorecards, scenarios and portfolios, integrated into more comprehensive schemes of time-to-market-management or strategic planning concerning business areas, and linked to elaborate procedures of computer based analysis and evaluation. In this respect, technology roadmapping enables a connection between tools, concepts and organisational units.

8 Recent Developments

Over approximately the past thirty years, technology roadmapping has gradually outgrown its infancy. Since it first emerged in the late 1970s, Roadmapping has developed into an acknowledged method of futurological research (see: Technology Futures Analysis Methods Working Group, 2004; Burmeister, Neef, 2005 including numerous examples from different industries). Moreover, it has become a stalwart tool of operational technology and innovation management (Moehrle, 2000, for the origins and early days of technology roadmapping see Probert, Radnor, 2003). In the course of its advancement, technology roadmapping has moved forward to various additional fields of application and generally expanded its scope of utility (see Bucher 2003 and DaCosta et al., 2003).

- In many fields of industry, technology roadmapping has by now come to be a time-tested practical instrument. It is, in fact, an acknowledged component of the elementary "toolbox" employed in different functional areas. Moreover, it is put to use in terms of cross-company coordination, e.g. in value-added chains or client-supplier-networks, and for the supervision of technological developments

in an intersectoral context as well as for the promotion of research. For instance, technology roadmaps for a planning interval of ten years' time were devised in the course of numerous so-called specific support actions of the European Commission's 6th General Research Programme, including the MONA-Project for the integration of optics and nanotechnology, and for areas of nanotechnology application (see Holtmannspötter et al., 2006, p. 221).
- In parallel with the establishment of technology roadmapping in companies and institutions of various dimensions and directions, a similar development has been observable regarding its scope of utility. Today, technology roadmapping is utilized in an increasingly wide range of planning activities and successfully combined with other tools and methods (see Phaal et al., 2004). However, this expansion to other fields of application and the increasing scope of utility also lead to higher expectations concerning the performance of technology roadmapping. Applications are getting more and more complex. Consequently, there is a growing necessity to support the utilization of technology roadmapping by means of modern information and communication technologies (ICT).

The demand for theoretical orientation as well as practical support in technology roadmapping continues - and continues to be strong. This has found corroboration in many talks between the editors of this book and representatives from science and industry. Technology managers, whether they are employed by industrial companies or academic institutions, long for a solid technology- and customer-related orientation and the suitable instruments for a direct operative implementation. And this is precisely what technology roadmapping has to offer.

References

Bucher, P.E.: Integrated Technology Roadmapping: Design and Implementation for Technology-Based Multinational Enterprises. Dissertation Thesis, Swiss Federal Institute of Technology Zurich (2003)

Burmeister, K., Neef, A. (eds.): In the Long Run. Corporate Foresight and Long Range Planning in Companies and Society. Oekom, Munich (2005)

Burgelman, R.A., Christensen, C.M., Wheelwright, S.M.: Strategic Management of Technology and Innovation, 4th edn. McGraw-Hill, Boston (2004)

Da Costa, O., Boden, M., Punie, Y., Zappacosta, M.: Wissenschafts- und Technologie-Roadmapping: Von der Industrie zur öffentlichen Politik. IPTS Report, 73 (2003)

European Industrial Research Management Association (ed.): Technology Roadmapping. Delivering Business Vision, Paris (1998)

Holtmannspötter, D., Rijkers-Defrasne, S., Glauner, C., Korte, S.: Aktuelle Technologieprognosen im internationalen Vergleich. Übersichtsstudie. Zukünftige Technologien Nr. 58. Düsseldorf: VDI Technologiezentrum (2006)

Moehrle, M.G.: Aktionsfelder einer betriebswirtschaftlichen Technologievorausschau. Industrie Management 16(5), 19–22 (2000)

Phaal, R., Farrukh, C.J.P., Probert, D.R.: Technology Roadmapping – A Planning Framework for Evolution and Evolution. Technological Forecasting & Social Change 71, 5–26 (2004)

Porter, A.L., Cunningham, S.W.: Tech Mining. Exploiting New Technologies for Competitive Advantage. Wiley, Hoboken (2005)

Probert, D., Radnor, M.: Frontier Experiences from Industry-Academia Consortia. Research Technology Management 42(2), 27–30 (2003)

Sematech, International Technology Roadmap for Semiconductors (2011), http://www.itrs.net/Links/2011ITRS/Home2011.html (accessed April 23, 2012)

Technology Futures Analysis Methods Working Group, Technology Futures Analysis: Toward Integration of the Field and New Methods. Technological Forecasting & Social Change 71, 287–303 (2004)

Authors

Martin G. Moehrle is director of the institute for project management and innovation (IPMI) at the University of Bremen, Germany, since 2001. From 1996 to 2001 he was leading the chair for planning and innovation management at the Technical University of Cottbus, Germany. The major research interests of IPMI are technology forecasting, evaluation of innovations, patent strategies, and TRIZ. Prof. Moehrle has published several books and articles in the field of innovation and technology management.

Ralf Isenmann joined the Fraunhofer Institute for Systems and Innovation Research ISI, Karlsruhe (Germany), in 2008. His research areas are in the interface between the management of innovation and technologies on the one hand and sustainable management, industrial ecology and corporate social responsibility (CSR) on the other. Special foci are corporate foresight projects and applications of technology roadmapping at corporate, national and international level for various industrial and governmental clients. He is an Associate Professor at the Faculty of Economics and Business Studies, University of Bremen. Since 2010 he is temporarily leading the Department for Sustainable Management at the Faculty of Business Studies, Management and Economics, University of Kassel.

Robert Phaal is a Principal Research Associate in the Engineering Department of the University of Cambridge, based in the Centre for Technology Management, Institute for Manufacturing. He conducts research in the area of strategic technology management, with particular interests in technology roadmapping and evaluation, emergence of technology-based industry and the development of practical management tools. Rob has a mechanical engineering background, with a PhD in computational mechanics, with industrial experience in technical consulting, contract research and software development.

Part 1: Institutional Reference for Technology Roadmapping

It appeared early on that technology roadmapping is a method that can be used in different contexts, at both company and industry levels, and within different management frames. Many industry-level roadmaps have been developed, providing useful guidance for related companies to structure common precompetitive R&D and competitive R&D programmes.

In part 1 of the book the institutional reference for technology roadmapping will be explored. First, technology roadmapping is a method for the strategic management of technology within a company. Technology roadmaps help as core documents for technology forecasting and evaluation, with technology roadmapping as a process for developing those documents. Technology roadmapping can also be seen as one of a set of methods for the management of network cooperation, helping the collaborating organizations to jointly define their aims and results and is also a tool to structure information in supply chains. Last not least, technology roadmapping can be applied on the state or public level, for instance to define coordinated lines of action within a specific field of technology.

- *Phaal, Farrukh and Probert* first gives an orientation about the process of technology roadmapping at the firm level. Building on definitions of 'technology' and 'technology management', the author presents a technology management framework that is used to position roadmapping in the business context. Then he shows how the framework could support communication and co-operative working across the organization as a key theme, with roadmaps providing a common structure and language for supporting innovation and strategy. The application of the theoretical framework is then illustrated by a case study describing how a medium-sized company implemented roadmapping over a period of several years.
- *Petrick* provides a substantial account of how technology roadmapping can be used at the level of supplier networks. The process of roadmapping enables participants to develop a shared understanding, to structure diverse information, supporting discussion and decision making. By means of easily comprehensible examples the author demonstrates how roadmaps support coordination, collaboration and cooperation in networked innovation systems. Firms acquire a better understanding of their position within the network, their contributions to the goals of that network and how they can influence their network and the environment by developing their own roadmapping strategy. The author concludes that it is particularly important for firms to understand that they compete against other networks, and that excellence of a single firm does not automatically lead to success of the network.

- *Nimmo* explores the application of the technology roadmapping process at the industry level, as supported by the Canadian government. This involves a description of the Canadian Technology Roadmap Model, which comprises three phases: development, implementation and evergreening. The experience of technology roadmapping in Canada is illustrated with two different examples; first a high-impact technology roadmap introduced by an industrial network, and second a roadmap created in conjunction with government institutions. The author explains how technology roadmapping is not primarily concerned with predicting future breakthroughs in science or technology, but rather emphasizes the process of roadmapping bringing together different parties, for example from industry and government, helping to analyse future challenges and opportunities.
- *Kerr, Phaal and Probert* focuses on technological roadmapping at the government level, interpreting it as a process for implementing government policy such as white and command papers, and for translating these documents into strategic plans. Using a case study, the author considers the future challenges and objectives of the Royal Australian Navy, prescribed by the government, and how this can be depicted visually in a roadmap. He explains in detail the individual steps of the process using a template. Roadmapping is presented as a tool which facilitates the decomposition of future visions into mayor projects which can transmitted in a political context.

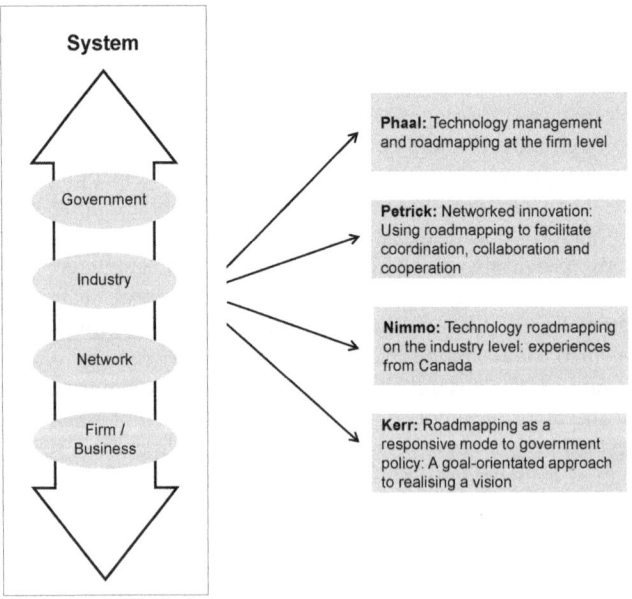

Fig. 1 Institutional references for technology roadmapping

Technology Management and Roadmapping at the Firm Level

Robert Phaal, Clare Farrukh, and David R. Probert

Technology is a key resource in many firms, as a basis for generating competitive advantage, and as an enabler for many activities. Technology management frameworks, processes and tools provide the means through which this valuable resource can be managed to ensure that technological investments are aligned with business goals and systems. In this chapter a practical framework for technology management is presented, highlighting five key processes: identification, selection, acquisition exploitation and protection. Technology roadmapping is proposed as a key integrating framework and management method that supports the implementation of these processes in firms, bringing together technical and commercial perspectives to support innovation and strategy.

1 Introduction

Technology has been a fundamental driver for innovation throughout the development of human society. With advances in fields such as information and communication technology, biotechnology and nanotechnology, the pace of innovation and change is set to increase further in the 21st century. This poses multiple challenges, for individuals, society and organisations, where managers are faced with hard decisions concerning how best to allocate limited resources, in terms of the increasing cost, complexity and risk of technology investments, against a background of increasing global competition. Technology management is a field of practice and academic enquiry that seeks to address these challenges. Within this field, technology roadmapping has emerged as a key method for linking technology decisions and investments to business objectives, supporting both strategy and innovation in the firm.

Technology management addresses the effective identification, selection, acquisition, exploitation and protection of technologies needed to maintain a

---

Robert Phaal · Clare Farrukh · David R. Probert
Centre for Technology Management, Institute for Manufacturing,
Department of Engineering, University of Cambridge, 17 Charles Babbage Road,
Cambridge, CB3 0FS, United Kingdom
e-mail: {rp108,cjp22,drp1001}@cam.ac.uk

stream of products and services to the market (Gregory, 1995). It deals with all aspects of integrating technological issues into business decision making and is directly relevant to a number of core business processes, including strategy development, innovation, new product development and operations management. Healthy technology management requires establishing appropriate knowledge flows between commercial and technological perspectives in the firm, to achieve a balance between market 'pull' and technology 'push'. The nature of these knowledge flows depends on both the internal and external context, including factors such as business aims, market dynamics and organisational culture.

The meaning of 'technology' and 'technology management' is explored in Section 2, as a basis for the technology management framework presented in Section 3. The practical application of the framework is demonstrated in Section 4, with reference to roadmapping, which is widely deployed within firms to support strategy and innovation, linking technological choices and investments to business and market objectives. A case study is presented in Section 5 to illustrate the application of roadmapping within a printing company, as a means for improved strategic technology management.

2 Technology and the Management of Technology

While there are many published definitions of 'technology' in the literature (for example, Steele, 1989; Whipp 1991), for practical purposes it can be defined as the *'know-how'* of the organisation. Technology, in the business context, can best be considered as an important type of resource, and hence there are considerable linkages with other resource-based views of the firm (Penrose, 1995; Grant, 1996), such as competence and capability approaches (Hamel and Prahalad, 1994; Teece 1980), and the general knowledge management literature. A key objective of technology management is to ensure that technological resources are effectively linked to business requirements, which is the focus of the technology management framework proposed in Section 3, and a key benefit of the technology roadmapping approach.

For the purposes of this paper the following definition is adopted, proposed by the European Institute of Technology and Innovation Management (EITIM, 2011): *"Technology management addresses the effective identification, selection, acquisition, development, exploitation and protection of technologies (product, process and infrastructural) needed to maintain a market position and business performance"*.

This definition highlights two important technology management themes:

- Establishing and maintaining the linkages between technological resources and company objectives is of vital importance and represents a continuing challenge for many firms. This requires effective communication and knowledge

management, supported by appropriate tools and processes. Of particular importance is the dialogue and understanding that needs to be established between the commercial and technological functions in the business.
- Effective technology management requires a number of management processes. These processes are not always very visible in firms, and are typically distributed within other business processes, such as strategy, innovation and operations.

The framework described in Section 3 is primarily intended to support technology management in the manufacturing sector, at the firm level (although, owing to the generic nature of the framework, it is considered likely to have broader application). To improve understanding of the framework it is important to define the system within which it applies, in the context of technology management.

The manufacturing business systems model that has been adopted is that used by the University of Cambridge Manufacturing Leaders' Programme (MLP), which forms the basis for a company audit (Hillier, 2001). The MLP model is built up in three stages, or levels:

- Level 1: a simple resource-based process view, where resources are identified as comprising people and facilities, which are combined with operational processes to transform inputs into required outputs. Based on the discussions above, the technology base of the firm can be considered to be a sub-set of these resources and processes.
- Level 2: expansion of the model to the firm level, defining the manufacturing business, in the context of the value chain that links suppliers to customers, highlighting a number of important business processes. These processes are strategy development, supply chain management and new product introduction, supplemented by supplier and customer development processes.
- Level 3: expansion of the model to include the business environment in which the firm operates: industry sectors, competitors and suppliers, current available technology, customers / consumers and liability, environment and economy. The broader trends that govern the evolution of this business environment are included in the model (such as industry, technology, general societal, political and economic trends).

This type of model defines the system within which technology management considerations can be explored. The importance of defining the system, including its boundaries, interfaces and elements, and the relationships between them, is supported by general systems theory (Jackson, 2000). The 'soft' systems perspective (Checkland, 1981), where the importance of how people perceive and interact with the system, is of particular relevance to technology management, which requires co-operation between technological and commercial functions (Linstone, 1999).

The concepts discussed in this section (technology as an important resource in the firm and the technology management processes that operate on the technology base, in the context of the manufacturing business system) provide the components on which the technology management framework is based.

3 Technology Management Framework

The overall aim of the framework (Fig. 1) is to support understanding of how technological and commercial knowledge combine to support strategy, innovation and operational processes in the firm, in the context of both the internal and external environment. The many particular activities and aims that are associated with technology management practice in firms depend on the particular context and objectives. Detailed frameworks have been developed to support decision-making and action in specific areas – for example, open innovation (Chesbrough, 2003) and technology intelligence (Kerr et al., 2006).

Technology management processes: At the heart of the framework is the technology base of the firm, which represents the technological knowledge, competences and capabilities that support the development and delivery of competitive products and services, and other organisational goals. Five technology management processes (ISAEP) operate on the technology base (Gregory, 1995), which combine to support the generation and exploitation of the firm's technology base:

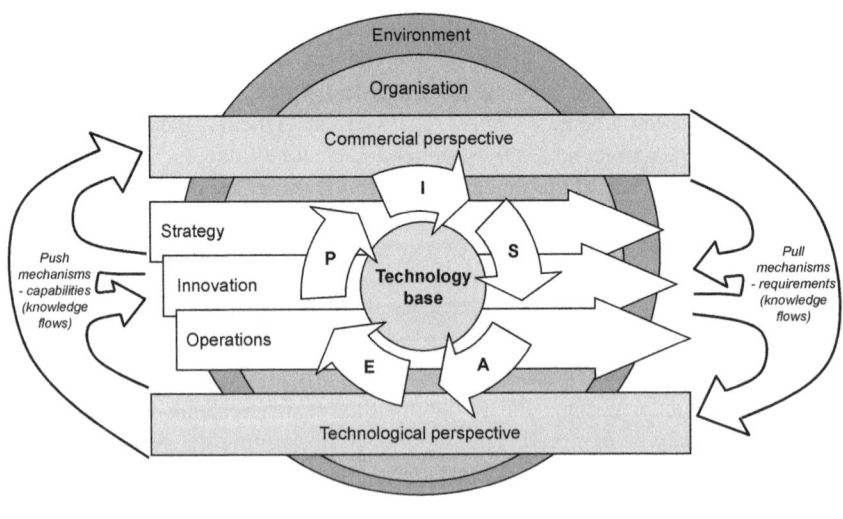

Fig. 1 Technology management framework

- *Identification* of technologies that are not currently part of the firm's technology base, but may be important in the future (for example, by attending conferences, reading journals, visiting trade fairs, questioning suppliers and conducting pure research).
- *Selection* of those technologies that the firm needs for its future products and technologies (for example, by using portfolio-type methods, expert judgement, pilot studies and financial methods).
- *Acquisition* of the technologies that have been selected (for example, by R&D, licensing, purchase of equipment, hiring of staff and acquisition of firms).
- *Exploitation* of the technologies that have been acquired (for example, by incorporating into products and services and licensing).
- *Protection* of the technological assets of the firm (for example, by legal means such as patenting, contracts, trademarks, copyright, together with security measures and retention of key staff).

Business processes: The ISAEP technology management processes do not operate in isolation, and are generally not managed as separate business processes. The various activities that constitute these management tasks tend to be distributed within other business processes (for instance, technology selection decisions are made during business strategy and new product development). Three core business processes are of particular importance: strategy, innovation and operations (SIO), operating at different business system levels in the firm. The link to core business processes is important, as these are the focus of management and action in the business and the means for ensuring sustainable productive output of the firm. The aim of effective technology management is to ensure that technological issues are incorporated appropriately into these processes, to form a technology management system that is coherent and integrated across and beyond specific business processes and activities.

Mechanisms for linking technological and commercial perspectives: The framework emphasises the dynamic nature of the knowledge flows that must occur between the commercial and technological functions in the firm if technology management is to be effective, linking to the strategy, innovation and operational processes. An appropriate balance must be struck between market 'pull' (requirements) and technology 'push' (capabilities). Various mechanisms can support the linkage of the commercial and technical perspectives, including traditional communication channels (for example, discussions and email), cross-functional teams / meetings, management tools, business processes, staff transfers and training.

Context: The specific technology management issues faced by firms depend on the context (internal and external), in terms of organisational structure, systems, infrastructure, culture and structure, and the particular business environment and challenges confronting the firm, which change over time.

Time: Time is a key dimension in technology management, in terms of synchronising technological developments and capabilities with business requirements, in the context of evolving markets, products and technology. Although time is not explicitly depicted in the framework, it is implicit in SIO business and ISAEP technology management processes.

The concept of 'pull' and 'push' mechanisms, which is a central feature of the technology management framework, is illustrated in Fig. 2 (Muller, 2000), which shows how 'mechanisms' such as people, information, documents, resources and processes connect key business processes, including commercial and technological perspectives.

As noted above, technology management processes (ISAEP) do not exist in isolation, and tend not to be managed as explicit core business processes, but rather are distributed as activities within other business processes, the most important of which are strategy, innovation and operations. Effective technology management requires that these relationships be understood and supported by effective knowledge management systems (pull and push mechanisms). The following points illustrate the complex relationships between these business and technology management processes:

First, the ISAEP technology management processes are not entirely linear in nature. While there is a logical flow from identification through selection, acquisition, exploitation and protection, some iteration and feedback is required. Broadly, the relationships between the processes can be described in terms of 'upstream' and 'downstream' flows. In the ISAEP sequence, information and knowledge generated during each activity can be useful for downstream processes. For example:

- Information gathered during identification of technology can be a useful input to the selection process. On the other hand, the identification of technology requires some form of 'filter' to direct efforts and to enable promising technologies to be recognised. This requires 'pre-selection', or a 'light' form of the selection process to be embedded within the identification process. Similar observations can be made for the other processes.
- In terms of feedback, each process can benefit from the learning that is generated by the application of the set of processes, which requires a systems-level perspective, with associated responsibilities for high-level technology management in the firm. The role of the technology management function includes the overall co-ordination of the activities that constitute the ISAEP processes and provision of the infrastructure for supporting their application, such as information and knowledge management, provision of management tools and training.

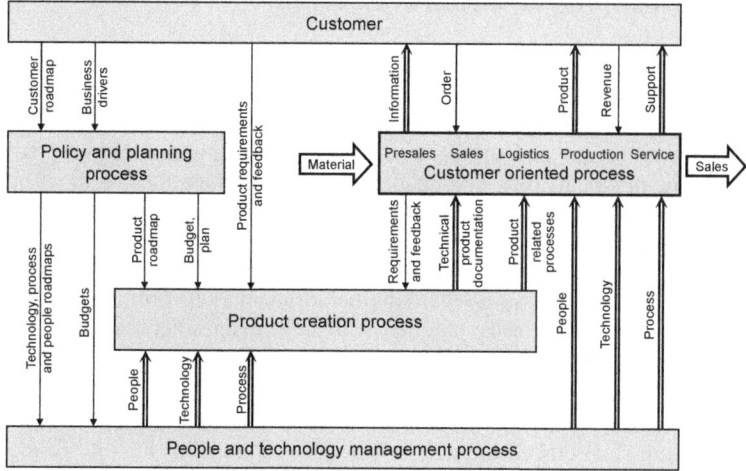

Fig. 2 Simplified decomposition of the business process (Muller, 2000)

Second, while the main focus of each SIO business process is at the corresponding SIO 'business system level', these processes also operate at the other levels. For instance:

- Strategy is primarily concerned with overall corporate or business objectives and direction, but also with innovation (for example, development of product and technology platforms) and operations (for example, how to best configure manufacturing and logistics).
- Innovation is primarily concerned with the development of new products and services, but also has a role to play in improving strategic and operations processes.
- Operations is primarily concerned with the flow of resources within the business system, but also with project management of the activities associated with strategy and innovation processes.

Thus, a complex picture emerges when assessing the relationship between the various ISAEP technology management and SIO business processes. Technological considerations impact on all of the business processes, at all levels of the business system. The processes for understanding and managing technology and the wider business are not simple or independent, but complex and intertwined. A holistic and predictive view of how all of these elements behave as an integrated system is perhaps too ambitious, owing to the context-dependent nature of many specific technology management tasks. The relative simplicity and generality of the technology management framework described in this paper encapsulates the principles that underpin effective technology management, across the breadth of the organisation and its activities.

4 Roadmapping at the Firm Level

Technology roadmapping has the potential for integrating processes and information across the span of the whole framework depicted in Fig. 1, and for supporting communication and co-operative working across the organisation.

The development and application of the roadmapping technique has been a focus of the practical work that underpins the technology management framework (Phaal et al., 2010). There are many types of roadmap, in terms of purpose and format, with the most common (generic) type shown in Fig. 3. The roadmap comprises a number of layers and sub-layers, within which the evolution or migration of the business is charted (including market, product and technology perspectives) on a time basis, together with key linkages between the layers.

Comparing the generic technology roadmap with that of the technology management framework (Section 3), it can be seen that there are some key structural relationships between the two, which highlight the importance of roadmapping for embedding the principles contained in the framework in industrial practice:

1. The commercial and technological layers of the roadmap directly relate to the commercial and technological perspectives in the framework and the linkages between these can be readily shown on the roadmap.
2. The linking (middle) layer of the roadmap (typically products or services, but more generally including other aspects such as business capabilities and systems) is closely related to the 'pull-push' linkage mechanisms in the framework. Generally, the middle layer of the roadmap can also be considered as a linkage mechanism, providing common ground for both the commercial and technological functions in the firm. For example, while staff in technology

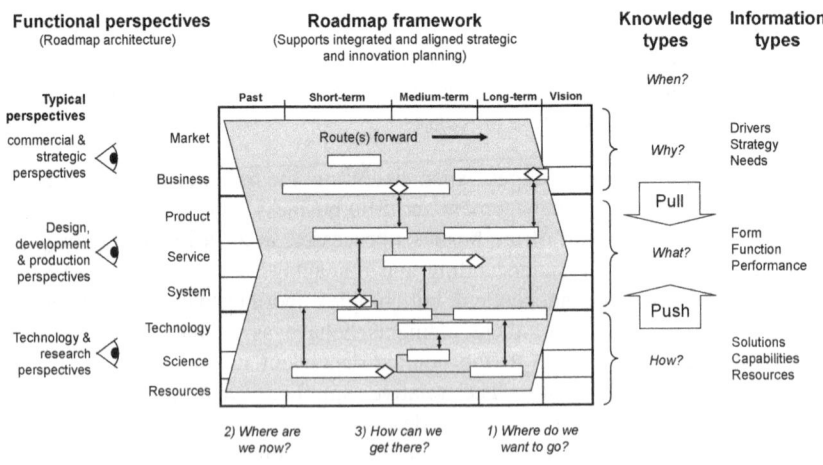

Fig. 3 Generalised technology roadmap structure

and marketing functions may approach the business from different perspectives, both groups have a sound understanding of products. The process whereby a roadmap is developed brings together representatives from all relevant functions, directly enhancing communication and co-operation.
3. The business processes (SIO) relate to the roadmap in a number of ways:
   - The operations, innovation and strategy processes are associated with different time horizons (short, medium and long, respectively), which are also closely related to the structure of the roadmap. The roadmap includes the temporal dimension explicitly, and tends to include short-, medium- and long-term perspectives, typically up to at least two innovation cycles into the future.
   - The strategy and innovation processes are often expressed at the business and product layers of the roadmap, respectively, in terms of strategic milestones, elements of strategy, new product introductions and service improvements. For roadmaps that are driven by technology push, elements of technology strategy may be incorporated into the technology layer.
   - The similarity of the roadmap structure to Gantt planning charts enables specific programmes and projects to be related directly to the roadmap, which are often used to monitor progress at a high level.

In summary, technology roadmaps support integrated technology management in business, owing to:

- The flexibility of roadmapping in terms of its application and structure (the approach can be used for supporting many planning-oriented activities, at any level in the firm).
- The close relationship between the structure and use of roadmaps and the business (SIO) and technology management (ISAEP) processes.
- The strength of technology roadmapping for supporting the linkages between commercial and technological perspectives in the firm.
- Many tools and techniques may be used for supporting strategy and planning (for example, competitor assessment, market research, technology audit and forecasting). The information that is generated using these approaches is a valuable input to the roadmapping process, which has the potential to act as a focal point for these activities.

Roadmapping is a scalable approach, enabled by the hierarchical taxonomy that defines the layered structure, and can be applied at various levels, ranging from products to firms and entire sectors. The original form of technology roadmapping, as developed by Motorola and other organisations focused on product-technology roadmapping – i.e. the technologies developments needed to realise a particular product strategy (Willyard and McClees, 1987; Phaal et al., 2001). More recently roadmapping has been applied at the firm level (business unit and corporate applications), applicable to a portfolio of products and business strategy in general (Cosner et al., 2007; Phaal et al., 2007).

The structure of roadmaps and the processes used for developing roadmaps need to be adapted to suit the particular purpose and organisational context. There is a close alignment between strategy and innovation processes and roadmapping, and often roadmapping is integrated into these processes, with roadmaps updated for key review and decision milestones. A generalised innovation/strategy process is often represented as a series of phases and review screens, associated with an iterative process that progresses and down-selects options, depicted schematically as a 'funnel' (see, for example, Cooper, 2001, 2006; Chesbrough, 2003). Roadmapping can be used throughout this process, although the content of the roadmap/s will change, and the process will be dramatically different on the left compared to the right. On the left the process will be inherently exploratory, helping to scope the challenge, identify, investigate and prioritise initial options to work on, progressing to a more controlled system on the right that shares some characteristics with project planning, at a strategic programme level, concerned with the delivery of complex projects.

An example is shown in Fig. 4 for Lucent Technologies (Albright and Kappel, 2003), which summarises the key steps associates with product-technology roadmapping. This involves completing a series of one-page templates covering various aspects that need to be included in a sound business case, starting with market considerations, then product and technology, leading to actions (a market pull approach). Notice that there are two roadmaps included in this process – a product roadmap and a technology roadmap. In contrast, the approach illustrated in Fig. 4 presents an integration of technology, product, market and other perspectives. This higher-level view was originally developed by Philips (Groenveld, 1997), and has the advantage that the important linkages between these perspectives are more clearly articulated.

Many variations of roadmapping have been developed by different organisations for different purposes. As a reference point, the approach proposed by the Sandia National Laboratories (Garcia and Bray, 1997) provides a useful basis for assessing and designing roadmapping processes – summarised in Table 1.

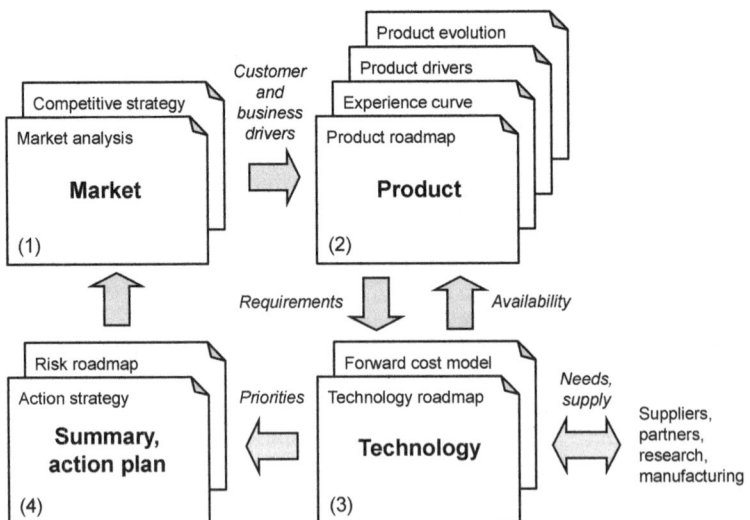

Fig. 4 Lucent Technologies roadmapping process (adapted from Albright & Kappel, 2003)

Table 1 Sandia National Laboratories' technology roadmapping process (Garcia and Bray, 1997)

Phase	Step	Notes
I. Preliminary activity	I.1 Satisfy essential conditions	Success factors that should be considered at the start of the process include: • Perceived need for roadmapping and collaborative development. • Input and participation from all relevant groups (for example, functions, customers, suppliers, partners, government agencies, universities). • Clarity of the boundaries of the initiative how roadmap will be used.
	I.2 Provide leadership / sponsorship	Committed leadership / sponsorship is needed due to the effort required in developing a roadmap and if it is to have impact, from key decision makers and those involved in implementing the roadmap.
	I.3 Define the scope and boundaries for the technology roadmap	The context of roadmap needs to be understood, including definition of the vision for the organisation, the aims of the roadmapping initiative, scope and boundaries, level of required detail, and timeframes.
II. Development of the technology roadmap	II.1 Identify the 'product' that will be the focus of the roadmap	The product needs and focus must be agreed if buy-in is to be achieved and sustained. Garcia & Bray recommend the use of scenario planning if there is major uncertainty about the project needs (see Chapter 5).
	II.2 Identify the critical system requirements and their targets	Critical system requirements need to be defined, including time-based targets (quantified if possible). These requirements relate to the functions and performance required from the product or system. These subsystems, functions and performance dimensions form core elements of the roadmap structure and process.
	II.3 Specify the major technology areas	These are the main technical areas that can contribute to the critical product or system requirements.
	II.4 Specify the technology drivers and their targets	The product or system requirements and targets need to be translated into technology drivers and targets for the major technology areas. These are criteria that can be used to evaluate the benefits of the technology, as a basis for differentiating the various options for selection purposes.
	II.5 Identify technology alternatives and their time lines	Technology alternatives should be identified, which have the potential to respond to the technology drivers and achieve the targets. Breakthroughs in several technologies may be required for challenging targets, and particular technologies might address multiple drivers. The timeframes in which the technology might mature sufficiently to be implemented within the products and systems must be estimated.
	II.6 Recommend the technology alternatives that should be pursued	The most attractive technologies need to be selected, which have the potential to achieve the desired targets, bearing in mind costs, development times, risks and the tradeoffs between these factors. Various tools and techniques may be helpful during this step, to support analysis and decision-making (see Chapter 5), although expert judgement is often a key factor, benefiting from a collaborative process. The output from this step is the graphical representation that is the focal point of the roadmap document or report.
	II.7 Create the technology roadmap report	The information generated from the above steps needs to be pulled together into an integrated report, including the graphical roadmap, description of each technology and its current status, critical risks and barriers, gaps, technical and implementation recommendations.
III. Follow-up activity	III.1 Critique and validate the roadmap	Development of the first (draft) version of the roadmap usually involves a relatively small group of key participants. Broader consultation is beneficial for validation purposes, to address key gaps identified and to build broader buy-in from those involved in or who influence its implementation. The roadmap should be updated as appropriate.
	III.2 Develop an implementation plan	If the roadmap is to have impact then the recommendations need to be implemented, which requires activities and projects to be planned, resourced, coordinated and managed.
	III.3 Review and update	Roadmaps should be reviewed and updated as appropriate, to reflect changing circumstances and learning. Typically this will be linked to business processes such as strategy and new product development or as events require.

5 Case Study – Aligning Technology and Product Developments

This case describes how a medium-sized company (1,500 employees) that develops and manufactures printing solutions for industrial applications implemented roadmapping over a period of several years (Phaal et al., 2008). The T-Plan method was used as a basis for this initiative, which brings together technical and commercial staff in a four half-day workshop process (Phaal, 2001a).

The business is organised primarily around four business units, each focusing on a different product line, with some overlap in technology and markets. The company headquarters are in Europe, co-located with core design and manufacturing operations, with regional centres and sales and support organisations based around the world, in more than 150 countries. The company is 30 years old, and has a strong technology heritage. As the company has grown in size and complexity, new technologies have been acquired and the product range expanded, with a need to establish methods to manage the effective acquisition and integration of technology into the core new product introduction process.

As a technology-based company, the firm was particularly aware that developing new technologies (or other competences) could take a long time. The company had had experience of including new technologies in product development projects before they were fully tried and tested. The result had always been delay and disappointment. To avoid this it was clear that they needed a coherent product-technology strategy so that innovations could be developed in advance and then brought to market quickly and securely when required, and roadmapping was selected as the most appropriate approach.

Roadmapping was first applied in the largest and oldest business unit, which is based on mature continuous inkjet printing technology. The main outcomes of this application were the recognition that too many projects were being pursued and that there was a lack of confidence that the market drivers were up-to-date. A market research study was undertaken, and the roadmap revised, and a series of new product initiatives followed. Based on this experience the method was rolled out to the other parts of the business.

Figure 5 shows an example of the first roadmap developed in one of the business units, forming the basis of an iterative process for reviewing and updating the roadmap on a regular basis. Figure 6 shows a more recent version of this roadmap, illustrating how the method evolved over a period of several years.

The roadmapping process was used in all business units in the firm, and proved an effective way to develop and articulate strategy efficiently and quickly. In all cases the first roadmap showed that the existing plans and intentions were too ambitious, and had to be scaled back – a valuable early result and a useful benefit from the work.

Technology Management and Roadmapping at the Firm Level

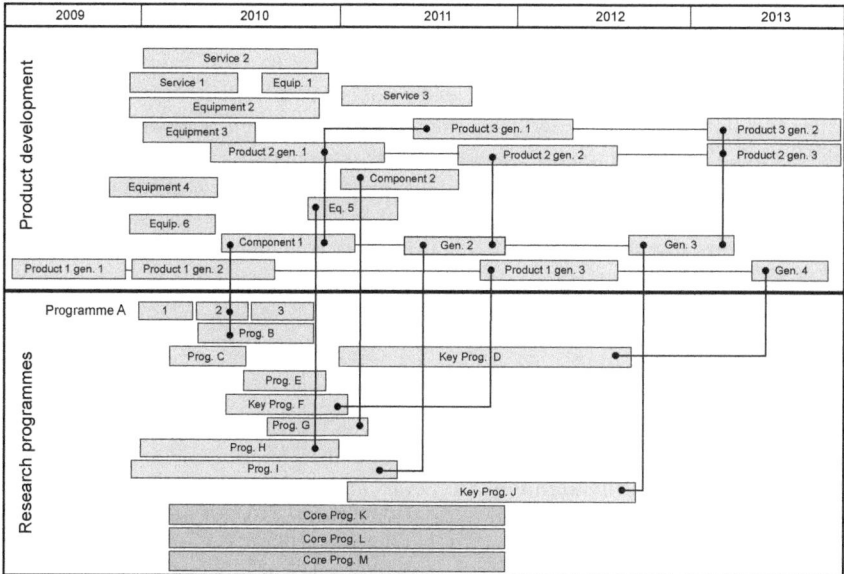

Fig. 5 Initial product-technology roadmap

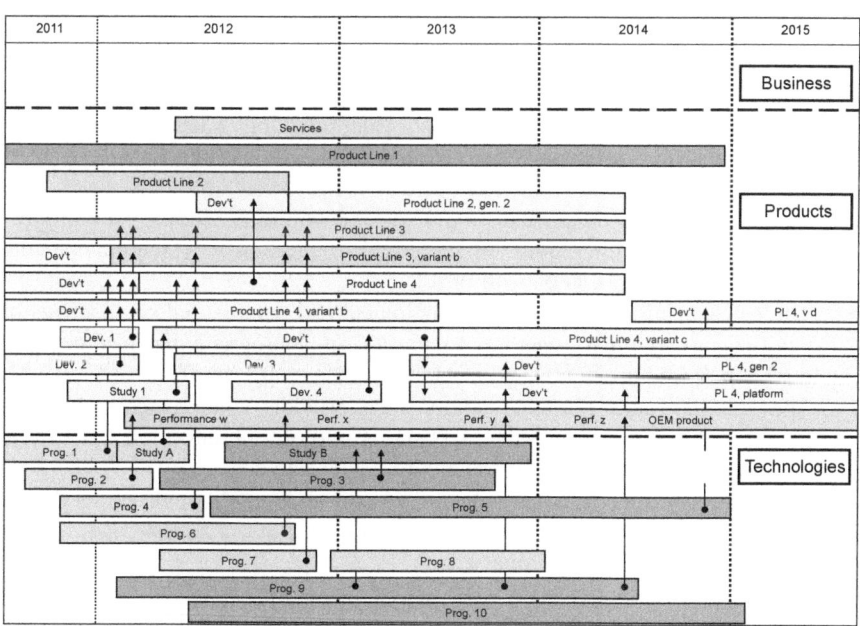

Fig. 6 Mature product-technology roadmap (project view)

Nevertheless, managers and staff usually treated the first versions of the roadmaps with caution and only really trusted them after they had been through several iterations. These reviews, typically every six months, were crucial. They gave time for participants to gather extra data and to reflect on what had been done. Inevitably the maps evolved and stabilised with repeated discussion, with the process of debate cementing understanding and support.

The roadmaps became a useful and valued tool for communicating the emerging strategy to the board and others in the company. Bringing the business unit roadmaps together helped to identify synergies that could lead to further efficiencies.

6 Summary

This paper has presented an overview of the development and application of a high-level framework for technology management. The framework is intended to be broad in scope, incorporating a number of key principles that underpin technology management. The framework, which is consistent with concepts from resource-based and systems thinking, provides a bridge between theory and practice.

The technology management framework incorporates the following key elements, which are all important for the understanding and application of technology management in business:

- The technology base of the firm, a key resource for many innovative companies.
- The technology management processes (identification, selection, acquisition, exploitation and protection) that operate on the technology base to support innovation in the firm.
- The core business processes of strategy, innovation (including new product development) and operations, which provide the means by which the potential value of technology can be realised.
- The mechanisms by which the technological and commercial perspectives of the firm are brought together, to ensure an appropriate balance between market pull (requirements) and technology push (capabilities).
- The internal and external factors that provide context to technology management in the firm, such as business purpose, organisational structure, culture and infrastructure, market environments and drivers.
- Time is a key dimension in technology management; although not explicitly depicted in the framework, it is implicit in SIO business and ISAEP technology management processes.

Technology roadmapping provides a means of addressing all of these challenges, with its use expanding both at the firm and sector levels as a core tool for supporting technology management. Its application to product-technology planning has been illustrated with an example for a firm in the printing sector, where the technique has been applied across the business to coordinate technology and commercial strategy. Owing to the context-dependent nature of particular technology management challenges, business processes and management tools need to be flexible, and be able to be adapted or customised to fit the aims, needs, resources and culture of the particular firm. The technology management framework and roadmapping approach both accommodate this requirement for flexibility.

References

Albright, R.E., Kappel, T.A.: Roadmapping in the corporation. Research Technology Management 42(2), 31–40 (2003)
Checkland, P.B.: Systems thinking, systems practice. Wiley, Chichester (1981)
Chesbrough, H.: Open innovation: the new imperative for creating and profiting from technology. Harvard Business Scholl Press, Boston (2003)
Cooper, R.G.: Winning at new products – accelerating the process from idea to launch, 3rd edn. Basic Books, New York (2001)
Cooper, R.G.: Managing technology development projects. Research Technology Management, 23–31 (November-December 2006)
Cosner, R.R., Hynds, E.J., Fusfeld, A.R., Loweth, C.V., Scouten, C., Albright, R.: Integrating roadmapping into technical planning. Research Technology Management, 31–48 (November-December 2007)
EITIM, European Institute of Technology Management (2011),
    http://www.eitim.org (accessed December 15, 2012)
Garcia, M.L., Bray, O.H.: Fundamentals of technology roadmapping. Report SAND97-0665. Sandia National Laboratories (1997)
Grant, R.M.: Toward a knowledge-based theory of the firm. Strategic Management Journal 17, 109–122 (1996)
Gregory, M.J.: Technology management – a process approach. Proceedings of the Institution of Mechanical Engineers 209, 347–356 (1995)
Groenveld, P.: Roadmapping integrates business and technology. Research-Technology Management 40(5), 48–55 (1997)
Hamel, G., Prahalad, C.K.: Competing for the future. Harvard Business School, Boston (1994)
Hillier, W.: The manufacturing business audit. Manufacturing Leaders Programme. University of Cambridge (2001)
Jackson, M.C.: Systems approaches to management. Kluwer Academic / Plenum Publishers, New York (2000)
Kerr, C.I.V., Mortara, L., Phaal, R., Probert, D.R.: A conceptual model for technology intelligence. International Journal of Technology Intelligence and Planning 2(1), 73–93 (2006)
Linstone, H.A.: Decision making for technology executives - using multiple perspectives to improve performance. Artech House, Boston (1999)
Muller, G.: Positioning the system architecture process. Philips Research (2000)
Penrose, E.: The theory of the growth of the firm. Oxford University Press, Oxford (1995)
Phaal, R., Farrukh, C., Probert, D.: Roadmapping for strategy and innovation – aligning technology and markets in a dynamic world. Institute for Manufacturing, University of Cambridge, Cambridge (2010)

Phaal, R., Mitchell, R., Probert, D.: T-Plan and S-Plan fast-start roadmapping approaches, Introduction and application of roadmapping. In: Khotsuki, S. (ed.) R&D Management, pp. 73–89. Technical Information Institute, Tokyo (2008)

Phaal, R., Farrukh, C.J.P., Probert, D.R.: Strategic roadmapping: a workshop-based approach for identifying and exploring innovation issues and opportunities. Engineering Management Journal 19(1), 16–24 (2007)

Phaal, R., Farrukh, C.J.P., Probert, D.R.: T-Plan: the fast-start to technology roadmapping - planning your route to success. Institute for Manufacturing, University of Cambridge, Cambridge (2001)

Steele, L.W.: Managing technology – the strategic view. McGraw-Hill, New York (1989)

Teece, D.J.: Economics of scope and the scope of the enterprise. Journal of Economic Behavior and Organization 1, 223–233 (1980)

Whipp, R.: Managing technological changes: opportunities and pitfalls. International Journal of Vehicle Design 12(5/6), 469–477 (1991)

Willyard, C.H., McClees, C.W.: Motorola's technology roadmap process. Research Management, 13–19 (September-October 1987)

Authors

Robert Phaal is a Principal Research Associate in the Engineering Department of the University of Cambridge, based in the Centre for Technology Management, Institute for Manufacturing. He conducts research in the area of strategic technology management, with particular interests in technology roadmapping and evaluation, emergence of technology-based industry and the development of practical management tools. Rob has a mechanical engineering background, with a PhD in computational mechanics, with industrial experience in technical consulting, contract research and software development.

Clare Farrukh is a Senior Research Associate in the Engineering Department of the University of Cambridge, based in the Centre for Technology Management, Institute for Manufacturing. Research interests include strategic technology management tools, new product introduction, technology valuation and industrial emergence. Clare has a chemical engineering background, with industrial experience in process plant and composites manufacturing, involving engineering projects, production support, process improvement and new product introduction work.

David Probert is a Reader in Technology Management and the Director of the Centre for Technology Management at the Engineering Department of the University of Cambridge. His current research interests are technology and innovation strategy, technology management processes, industrial sustainability and make or buy, technology acquisition and software sourcing. David pursued an industrial career with in the food, clothing and electronics sectors for 18 years before returning to Cambridge in 1991.

Networked Innovation: Using Roadmapping to Facilitate Coordination, Collaboration and Cooperation

Irene J. Petrick

The nature of relationships between firms has changed from a supply chain to a supply network and increasingly toward an ecosystem. The demands of each of these are different, as are the resulting uncertainties. Roadmapping is a method that at its core is used to buy down uncertainty. To achieve value the firm must understand its role in the network, and the way that its goals are related to the network's overall effectiveness. There are three aspects of the interaction between firms that are relevant to networked innovation: coordination (linking activity to time in a transaction-based relationship), collaboration (linking activity to intent in a co-creation relationship) and cooperation (linking activity to value creation across an ecosystem or platform). Ultimately, the way that roadmapping is approached should balance the goals of the firm and the network, the sources of uncertainty and the most likely types of interactions.

1 Introduction

Increasing product and service complexity suggests that few firms possess the entire production capability or knowledge base needed to design, manufacture and distribute most products or services. Instead, suppliers within networks add value to one another's activities. These networks of competing firms find their competitive advantage in both their product and/or service offerings *and* in their ability to align their decisions and activities with one another in more effective ways relative to other networks. In supply chain management, researchers have long understood the importance of the efficient transfer of goods and materials across participating firms. What is less well understood are management practices that might promote the rapid recognition of emerging opportunities and the integration and leverage of inter-firm knowledge networks (Stock, Boyer and

---

Irene J. Petrick
College of Information Sciences and Technology, The Pennsylvania State University,
102T Information Sciences and Technology Building, University Park,
PA 16802-6823, United States of America
e-mail: ipetrick@ist.psu.edu

Harmon, 2009). This is particularly important in firms that pursue an externally oriented innovation strategy where partnerships, alliances and collaboration are essential ingredients (Wunker and Pohle, 2007). Success in interfirm networks is achieved through access to information, resources, market insights and technologies (Gulati, Nohria and Zaheer, 2000) where inter-organizational learning (Hult, Ketchen and Nichols, 2003) and adaptation (Pathak, Day, Nair, Sawaya and Kristal, 2007) are continuous and robust practices.

This chapter identifies ways that roadmapping can be an important practice that facilitates inter-organizational network effectiveness. The discussion begins with the notion that roadmapping is an information organizing framework that helps create a shared understanding among participants. In networked innovation, this shared understanding then guides the choices that individual firms make about what and how to execute and when to deliver. There are three types of interactions that firms working within a supply network must address: (1) coordination – linking activity to time in a transaction-based relationship; (2) collaboration – linking activity to intent in a co-creation relationship; and (3) cooperation – linking activity to value creation across an ecosystem or platform relationship. As part of the discussion, I introduce the concept of supplier networks as complex adaptive systems. I suggest that the approach to roadmapping should be matched to the sources of the uncertainty and to the resultant needs for interaction between firms.

2 Conversations in the New Product/Service Development Space

Roadmapping has long been associated with *technology roadmapping* as a way to refine technology strategy and to align technology development with the critical path of a particular product launch (Kostoff and Schaller, 2001; Phaal, Farrukh & Probert, 2004). But this is only one of the conversations that must occur in the context of product or service innovation. In fact there are actually three separate conversations that should take place within a firm and for networked innovations, across firms, to envision and execute successful innovations around products and services. These three are conversations about:

- *Opportunity:* What are all the things that we might do?
- *Possibility:* What are all the ways that these things might be done? and
- *Action:* What will we actually do?

In essence, these three conversations reflect market pull, technology push and execution. Often, firms do not have these three conversations with the same degree of fervor, and the most time is spent in the *Action* discussion – what features should the product or service have; how will these features be reduced to requirements definitions; what is the timeframe in which this product or service will be launched; what are the underlying technical challenges we will have in bringing this product or service to market; and how will this actually happen? While these questions are all very critical to the ultimate success of the product or service, they beg the more important issues, namely what problem is this product or service intended to solve and how will it achieve this in a way that is obviously superior to other options that the customer or end user may have? Is our intended new product or service the best alternative to pursue? A second issue that gets even less attention is how the collective resources and knowledge of the supply network can be leveraged to achieve this superior product or service in a way that is inimitable by other networks of firms. Here the solution space should be explored more thoroughly to better understand alternatives to the anticipated technical solution.

2.1 Opportunity Conversations

As part of nearly any discussion about innovation, the participants are urged to think "out of the box". While this can result in unanticipated innovations, the true challenge is to redraw the boundaries of the collective box. Often this begins with a deeper understanding of that proverbial box by delving into the knowledge that is distributed within the network. To truly delight the market, the entire supplier network must come together to create compelling and unexpected products that immediately address an unmet need that the customer or end user did not even recognize. This requires a robust information gathering effort – particularly around users and the trends that drive their expectations. These nuggets of information are distributed across the supply network and are often tied to the domains or geographies in which the individual firms work. For example, in Figure 1 we can see that four facts when taken separately appear to offer little guidance in the way of specific product or service opportunities. When taken together, however, these facts combine to suggest a potential product and service opportunity, namely wirelessly enabled health monitoring clothing.

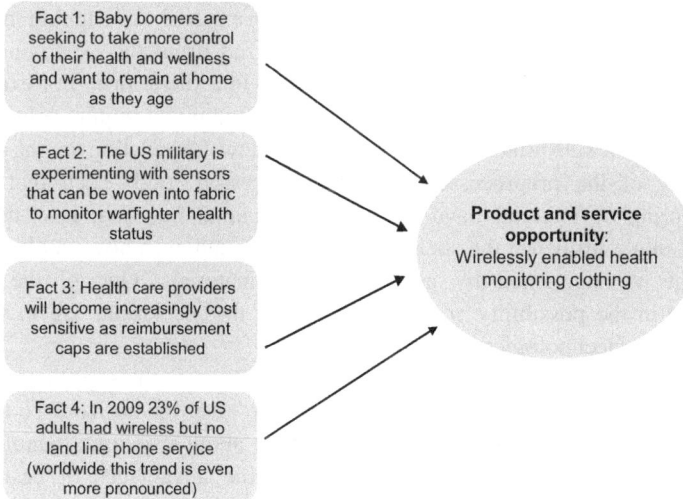

Fig. 1 Linking facts to identify non-obvious product and service opportunities

As firms seek to be more competitive globally, sensing differences between global consumers and their wants and needs will require insights that can only be gained by those within the network who are operating in that locale. For example, it took Tata Motors, an Indian firm, to understand the opportunity for its Nano car. Priced at a lowly $2,500 when it was introduced, the Nano was just what the Indian market needed – a car designed to fit easily on the streets of the most crowded cities, that was designed in a modular way so that it could be shipped in kits and then assembled by third parties if needed, and that was really little more than an enlarged motorbike with a fixed weather shield. The Nano is a complete departure from the Western automobile. But it took the combination of several facts to become a reality: the Indian worker spends a large amount of time commuting to and from their workplace; the Indian weather is predictably wet for long and intense periods of the year; there is a growing middle class in India that wants and can now afford transportation solutions; the maximum speed on Indian roads is less that the maximum speed on Western roads; the Indian transportation infrastructure leaves little space for parking, which is at a premium; and the Indian consumer does not have the same expectation for standard features as their Western counterpart, thus eliminating the need for higher priced items like air conditioning and power brakes in the basic model. Insight for new product or service opportunities will increasingly come from global sources.

2.2 Possibility Conversations

Possibility conversations are almost always about *how* something might be done and generally fall to the technologists. Here is where networked innovation is even more important since so much of the understanding of the intricacies of one technology solution versus another lies with the experts. As supplier networks have become populated with increasingly specialized firms, the experts needed to understand the technology landscape rarely reside in a single firm. For example, a food company seeking to create a healthier snack food might require solutions that include alternate substitutes for sugar, how the new formulation might alter the consistency of the preprocessed and then processed foods such that current manufacturing or material handling processes cannot be used, how the sugar substitute might influence the shelf life of the food, and even how an altered shelf life might be augmented by new packaging materials. One of the biggest challenges in the possibility conversation is that most solutions are in reality a combination of technologies.

Another important aspect of distributed expertise across the supplier network is understanding what technology substitutes are just "over the horizon" so to speak. For example, a car company may not be thinking about substitute manufacturing processes, but its suppliers may be weighing the advantages of casting and machining versus powder metals manufacturing as they continue to push for lower cost and near net shape manufacturing options. Or their suppliers may be considering how to reduce weight by using novel materials that may then require alternative manufacturing processes. The car company benefits when it taps into this supplier's knowledge.

2.3 Action Conversations

As noted earlier, action conversations are the most prevalent and get the most attention. It is the action conversation that drives individual firm planning within the larger networked innovation effort. Ideally, once the possibility and opportunity conversations have occurred, it is time to set common goals, articulate these common goals in terms of objectives for individual firms, and then to determine the timing of the various activities and tasks that must be undertaken. Action can be expressed as project plans, GANT or PERT charts, or through other execution management tools. This chapter is less concerned with the action that individual firms execute and is more interested in the way that this action gets initiated through the collective efforts of the network.

3 Roadmapping as an Information Organizing Framework

Within the context of these three conversations, roadmapping has a unique potential to support the underlying discussions because it offers the team a way to capture diverse information in a structured form. Figure 2 demonstrates the diversity of the information streams that need to be considered in the context of networked innovation where the capabilities of the supplier network are brought to bear. The goal of any roadmapping exercise should be to capture the external uncertainties within the market space and within the broader socioeconomic spectrum.

The capabilities that the supplier network possesses are rooted in the manufacturing processes and intellectual property strategies, as well as in the expertise resident within each firm across multiple functional areas. With respect to the product or service portfolio offerings, the competitive advantage that the network possesses lies in the collective execution across technology creation, manufacturing and intellectual property, and in the way that these are embedded into the product or service offered into the marketplace.

For both new product and service innovations, success comes to the firms and networks that can manage and/or leverage uncertainty. *How* uncertainty is managed and leveraged is directly related to the sources of the uncertainty in the first place. At its core, roadmapping is a method that is used to buy-down uncertainty by creating a shared understanding of the overall context and execution needs. Because of this roadmapping can be a very effective method when applied to the complexities of interfirm interactions needed to conceive

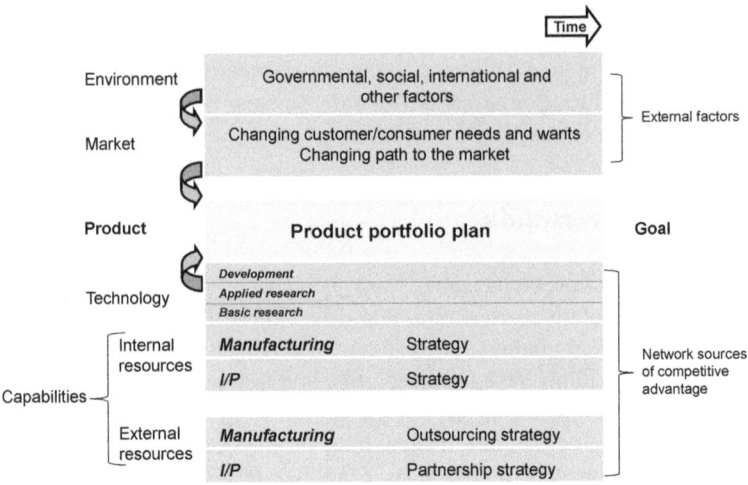

Fig. 2 The complexities of networked innovation

of, launch and then deliver new products and services. The benefits of roadmapping have long been understood to include the alignment of actions across functional areas within a firm as activities are sequenced in time and as their interdependencies are established. Such *a priori* views help create a shared set of objectives that can then be translated into individual task assignments and activities across functional areas. As the locus of innovation for many products and services is shifting to suppliers, the benefits to roadmapping will be constrained for those firms that do not take a supplier network approach to this effort.

Each firm that undertakes roadmapping in the context of networked innovation needs to consider the types of supplier relationships that will be needed to effectively and efficiently bring a new product or service to market. In networked innovations the nature of suppliers is strongly interconnected. In many cases we see that one or more firms within the network need to set direction about the product/service portfolio. From this a make versus buy decision must be made at multiple points within the supplier network. For those networks with distributed expertise, multiple roadmaps will need to be harmonized for success.

Pathik et al. (2007) note that supply networks are being forced to take into account a growing diversity of information available both from the surrounding environment and from increasing and evolving supply network partners. In this sense, the challenges of supplier interactions far exceed the traditional view of the supply chain. Thus it is useful to link the conversations in the new product/service and roadmapping with one final piece of the puzzle – the firm interactions that actually produce harmonized action.

4 Coordination, Collaboration and Cooperation in Networked Innovation

As individual firms have concentrated on their core competencies, the creation, manufacturing and delivery of goods and services has become an effort that can only be achieved through the combined actions of multiple firms. The nature of the interactions between firms acting in a network is contextual and can be described through an Interaction Continuum (see Table 1). Thompson (1967) first described task interdependency where he identified pooled dependency where different units use common resources but are independent; sequential dependency, where the output from one unit becomes the input to another unit and reciprocal dependency where units feed work back and forth. The ability of one unit to impact another unit is directly tied to the type of interdependence. From a supplier network perspective, both sequential and reciprocal dependencies introduce risks and uncertainties into underlying firm activities.

4.1 Coordination

The simplest form of interaction is *coordination* which is concerned with harmonizing tasks between more than one individual or between groups or firms. The intent is to avoid gaps and overlaps in assigned work and to enable the efficient transfer of outputs – in the case of a supply chain, the efficient and timely transfer of raw materials and components from one firm in the supply chain to another for further processing, assembly or distribution. In this sense, coordination is the act of managing interdependencies between activities performed to achieve an established goal. Here the interdependencies are understood, and the mechanisms used to achieve coordination can be a simple as establishing differing roles and responsibilities within the firm. Between firms, coordination happens through transaction-based contracts, and firms act as free agents in the marketplace.

Within the supply chain, coordination is most often used to describe a linear buyer-supplier dyad that has sparse connectivity. These dyads operate in a static environment and where the buyer and supplier exhibit fixed and non-adaptive behavior (Pathak et al., 2007). In the context of new product development activities, coordination also occurs through well established processes such as project management and critical path management. Within the coordination realm, the individual, group or firm has little latitude in carrying out the agreed upon task in the agreed upon timeframe, and the interaction is more like a handoff between actors in a well-scripted activity. In new product or service development, coordination is focused on managing the timing of activities between firms.

A good example of innovation that is the result of coordination is Motorola's interaction with its battery suppliers in the early 2000s. As Motorola missed a critical new cell phone introduction window due to a lack of sufficient batteries for its offerings, given demand volume, the company sought to identify key suppliers that it could work with to guarantee the needed volume of future batteries. As part of their efforts, Motorola actually provided key suppliers with additional capital to build molds to facilitate a higher production volume. This type of interaction required Motorola to share information about its intended production needs and to encourage its designers to reuse battery designs to insure enough demand to create the battery production volumes within its suppliers to achieve desired price points. In this example, Motorola and its key battery suppliers used advanced information to help stabilize the production and delivery schedule from the supplier to meet Motorola's anticipated needs. To achieve this, Motorola was willing to change its approach to purchasing to go from a strict per part cost to a pre-delivery expense to the battery supplier firm. Here, the coordination is more than a simple contract for delivery, and implies a commitment between Motorola and its key suppliers to facilitate product introduction.

4.2 Collaboration

Moving through the Interaction Continuum, *collaboration* requires a systems view to solve problems in a complicated environment where firms, often referred to as agents, obtain mutual benefit through their interaction. Here the collaborating agents have one or more shared goals and because of this, develop shared objectives that guide their actions. Collaboration requires mutual trust and generally results in the partitioning of tasks across individuals and groups where individual actors have considerable leeway in how these tasks are carried out. The principles guiding agent choices are rooted in these shared goals and objectives.

In the context of new product or service development within the supply network, collaboration results not only in efficiencies of activity, but in a savings of time and resources through the elimination of redundant activities. In addition, collaboration can result in novel approaches to solving problems as knowledge is shared through frequent consultation across a broad set of agents, many of whom possess unique expertise and capabilities (Petrick and Pogrebnyakov, 2005). Collaboration occurs as firms within the supply network seek to establish objectives, articulate needed action and match the timing of the actions with the shared goals of the network.

The Boeing Commercial Aircraft (BCA) 787 program is a very good example of collaboration. In its conceptual design, BCA chose to replace its traditional metal shell with composite materials while also replacing hydraulic controls with electronics. Because of this, BCA needed to establish integrated design procedures that enabled its suppliers to translate these high level concepts into working components, subassemblies and assemblies. BCA's intent was to leverage materials and electronics knowledge within its key supplier firms to co-create innovative solutions. Here, BCA was interested in both the timing of delivery and in the actual innovation contained in its suppliers' products. The 787 represents the capabilities that are distributed across BCA's 787 supplier network. Despite the delays, the 787 is likely to be Boeing's most successful commercial aircraft offering to date if the preproduction orders are any indication. Figure 3 demonstrates a portion of Boeing's 787 supplier network. Here, the careful reader will note both the dyad and triad relationships that helped support innovation for the 787.

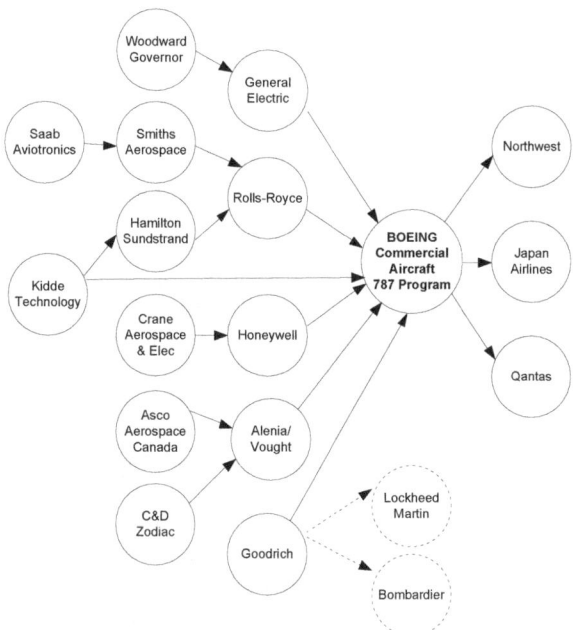

Fig. 3 Portion of the Boeing 787 Dreamliner supplier network

Another example of collaboration is General Motors (GM) use of Dupont's lightweight plastic Zytel® PLUS nylon for its new 2.0-litre Turbo engine. Here, the innovation is the result of new materials developed at a GM supplier and translated into a solution by a second GM supplier, Camoplast Polymer Solutions, located in Quebec, Canada. This winning combination was recently recognized by the Society of Plastics Engineers "Most Innovative Use of Plastics" 2010 award. One of the most interesting aspects of this example is that the new Dupont materials were only introduced in March 2010, and 90 days later are part of a qualified solution for the acoustic cover on the new engines. Here, collaboration between designers, manufacturers and material specialists within the three firms clearly resulted in the co-creation of an innovative solution to the high temperature and lower weight needs of the new engine.

4.3 Cooperation

Anchoring the far right end of the Interaction Continuum is cooperation where goals are jointly established and then evolve through the actions of the agents interacting. It should be noted that the terms *collaboration* and *cooperation* are often used interchangeably in the literature. In this chapter, I distinguish between the two using the theory of complex adaptive systems (CAS). CAS was first applied to supplier networks by Choi, Dooley and Rungtusanathan (2001) to describe a non-linear and dynamic environment in which the participant firms co-evolve with their environment. Here the relationships of the various agents cannot always be anticipated in advance and there is a high degree of situated decision-making that is context and time specific. Agents participating at one time period may be very different than those participating in a future time period, and in fact, these agents are responding to each other and to their environment. The experience that the collective participants have feeds forward over time to influence future activities and possibilities, often altering the scope or nature of the future environment. In this sense, cooperative interactions produce emergent behavior (Mintzberg and Waters, 1985).

The feed forward aspect of complex adaptive systems in the context of the supply network is a clear differentiating factor between collaboration and cooperation. Whereas in collaboration, firms work together to achieve joint goals, in cooperation, firms are working together not only to articulate joint goals, but to also proactively influence their future environment. These types of interactions are often described as occurring within an ecosystem (Iansiti and Levien, 2004).

There are several good examples of innovation that occurs as the result of cooperation within an ecosystem, including the iPhone and Facebook ecosystems. In both of these cases, the platform for development – the iPhone and its operating system on the one hand, and the social networking site's infrastructure on the other hand – offered developers the opportunity to leverage the platform to launch their own innovations and products. The creation of applications (known as apps) for both of these platforms increases the value of the platform. As more users flock to these platforms, there is an incentive for developers to create even more apps. This value creation based on network effects has been modeled theoretically by Metcalf's Law and has been described as a PIVA – platform of increasing value add (Gawer and Cusumano, 2002). To develop a successful PIVA, the creating platform firm must anticipate evolving future spaces where new combinations of technology and business models can create increased customer value.

Platforms such as Facebook and the iPhone are leveraging the activities of far flung developers. These developers are both users and creators, and the platforms leverage their talents and their insights to create ever more targeted and diverse applications. Once a successful PIVA is established it is very difficult for firms to compete against them. According to David Kirkpatrick (2010), Facebook has more than 500,000 applications operating on its platform, with one million registered developers located in more than 180 countries. In addition, more than 250 of these applications have at least one million active users. In fact the success of these and other platforms has led to the dominance of a single firm and its supporting ecosystem in several key sectors. Wu (2010) notes that Facebook owns social networking and Apple, maker of the iPod, iPhone and iPad and creator of iTunes owns on-line content delivery.

In the more traditional product development world, cooperative interactions can be seen in the move toward open innovation (Chesbrough, 2006). Here firms such as Toyota and Procter & Gamble have a stated goal of basing up to half of their innovations on ideas that come from outside of their traditional pipelines. The open innovation movement is growing, and a key challenge for these firms is attracting the mindshare of potential developers to focus on specific problems of interest to the company. This will lead to increasingly more dynamic ecosystems over time.

Table 1 The interaction continuum

	Coordination	Collaboration	Cooperation
Structure			
Agents	Dyads	Network	Ecosystem
Purpose	Task harmonization	Translation of shared goals into individual agent articulation of objectives and activities to facilitate co-creation	Strategic initiatives within and between agents to create value and proactively influence strategic future space
Mechanisms	Contractually established schedule, cost and performance parameters	Integrated design and development	Platform approach
Innovation Relevance	Timing	Activity planning to share capabilities and knowledge; Alignment of timing and approach across agents	Creation of new markets through unanticipated combinations of technology, business models, and products from diverse agents
Examples	Timing of delivery of a component or subassembly; Technology readiness appropriate for product launch	New material advances integrated into a new technology solution with design modifications to the product	Platforms of increasing value add (PIVA) where network effects drive value;

5 Addressing the Interaction Continuum in Roadmapping

In my experience, firms tend to believe that a single approach to roadmapping is sufficient for most of their innovation needs. This fallacy results in roadmapping efforts that fall short of hoped for benefits or that become a one-off with little sustainable value. Instead, firms should establish their roadmapping practices in the context of the particular new product or service development goals and in consideration for the role they play with respect to the network. Goals can be, and often are, established by the lead firm in the supplier network which has traditionally been perceived to be the Original Equipment Manufacturer (OEM). In recent years, these OEM's have recognized the need to work with their key customers or suppliers to jointly envision and articulate these goals. Some example goals include:

- To leverage supplier innovations around materials and manufacturing processes to create a lighter weight solution (for an engine, an aircraft fuselage, an automobile chassis, etc.)
- To provide a food product that has a guaranteed freshness that consumers can recognize with smart packaging that uses sensors to monitor important environmental aspects (such as vibration, temperature, moisture, etc.)

- To develop and deliver a smart device that can automatically sense location (via GPS) and match location to nearby facilities and destination spots (such as restaurants, hospitals, shopping areas) and wirelessly offer coupons or other enticements to device holders

Additionally, supplier firms with key intellectual property or with key capabilities have also been able to establish and drive these new product or service goals. Companies like Intel which provide key enabling technologies for many products have been able to effectively drive their ecosystem for more than two decades (Burgelman, 2002; Gawer and Cusumano, 2002).

Regardless where the driving goal is established, once it is established, firms within the network must consider alternative paths to market, alternative technology solutions, and the diversity of the various global consumers and their operating contexts. Here is where roadmapping can facilitate these discussions. But a single roadmap is inadequate to capture the nuances of the interactions.

5.1 Roadmap Sets

In point of fact, many firms traditionally envision what I call the composite roadmap when they think about roadmapping. The composite roadmap (See Figure 4) has stacked horizontal swim lanes that include the high level elements that must be aligned to successfully launch a new product or service. This is the map that is intended to drive action across multiple functions and often across multiple organizations. In this example, several product or service offerings are planned (from P1 to P5) to be sequentially offered into two different markets. In its best case, this product or service portfolio takes into account both market pull and technology push. To support these new offerings, Technology Platforms will need to evolve from T1 through T3 for example, and these in turn are supported by basic and applied research efforts within the lead firm and within its supplier network. Here, we can see that often the make versus buy decision can be planned in advance to anticipate potential failure of a hoped for breakthrough within the lead firm, where BR4a has a backup plan of a substitute supplier innovation achieved through BR4b and AR3.

Behind this composite roadmap for action, however, should sit assessments roadmaps for both the market and the technology solution space. Here is where various roadmaps can support the types of interactions that are needed to leverage the collective knowledge of the network and achieve new product or service innovations that will truly delight the market. The match of roadmap form and interaction type is detailed below. Not surprisingly, as we move from interactions around coordination to those of collaboration and cooperation, the deterministic nature of the composite roadmap needs to be augmented by environmental scanning and assessment efforts.

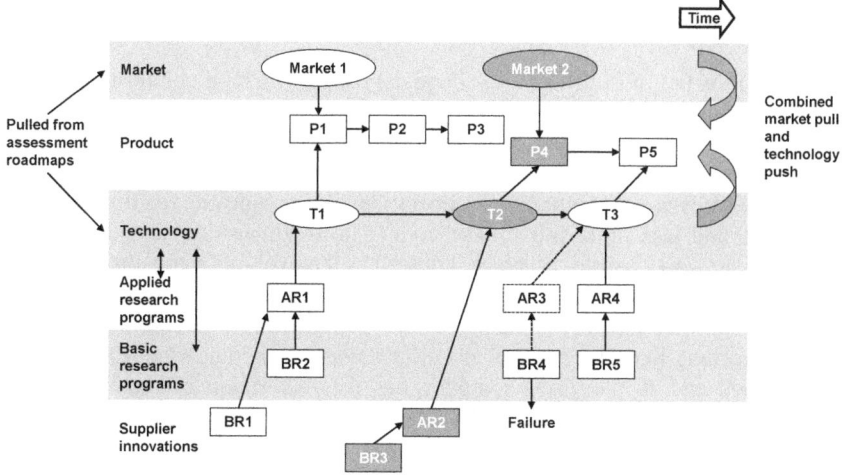

Fig. 4 A composite roadmap view of networked innovation for a new product or service

5.2 Coordination

Because coordination focuses on linking activity to time in a transaction based relationship, the composite roadmap is useful for alignment across functions within the lead firm or across key supplier firms. For example, as part of Motorola's battery initiative described earlier, Motorola and its suppliers jointly undertook roadmapping, feeding each other's maps, to share information about production volumes, feature needs and delivery timing based on supplier technology developments and Motorola's feature set needs and anticipated volumes.

5.3 Collaboration

Collaboration emphasizes linking activity to intent in a co-creation relationship between firms in the network. Here composite roadmaps can support technology planning when there are alternatives that might be used to create a successful technical solution. Such was the case in the Figure 4 example, where the lead firm hoped to be able to create its own intellectual property, but had a supplier alternative identified in case of failure.

In the Boeing example, assessing the technology landscape became a very important aspect of the 787's planning because much of the technology expertise needed to create the composite and electrical solutions needed for the new airplane design rested in the supplier base. Boeing actually created collaboration roadmaps to help support its selection of key design partners. These collaboration roadmaps facilitated the sourcing of raw materials such as titanium and identified key manufacturers who were globally distributed to help design, develop and the manufacture critical structures.

5.4 Cooperation

Cooperation is the most difficult to adequately support with a composite roadmap at the planning stages because the hallmark of cooperation is the coevolution of the network and the environment. Because of this, a composite roadmap can guide the platform creator's activities, but the timing is much less strict and the fluidity of the events is much higher. In cooperation, the composite roadmap is more descriptive and less proscriptive. The composite roadmap does not drive action among the network firms as much as it describes the evolving interactions.

Instead, the challenges of ecosystem and value creation innovation favor a robust market assessment roadmap that has inputs from a diverse set of agents who are working in niche markets and who possess niche knowledge. In the case of Facebook and Apple, each company has its own vision of the future. This vision of the future includes both the sourcing and creating of technology innovations and solutions, and so technology assessment remains an important consideration. However, the path forward in terms of features that end users see is determined by the apps that are created and the evolving way that users adopt these apps into new and unanticipated uses.

6 Conclusion

In a recent Bloomberg BusinessWeek Viewpoint, Dev Patnaik (2010) asserts

> "Innovation is about growth and growth takes empathy, creativity, and execution. Empathy, on an organizational scale, is a shared intuition for what people outside the company really need and value. Creativity is the ability to come up with new ideas for products, services, and businesses that are different and distinct. And execution is the art of getting things done. These aren't feel-good ideas for easy times. They're the secret to surviving the storm."

As firms face increasing competition and as networks of firms compete against other networks, excellence in execution by a single firm is no longer a guarantee of its success. Instead, each firm must understand its position within the network, its contribution to the overall goals of that network, and its ability to influence its network and environment (or not). When used properly, roadmapping is a framework that helps support the conversations that must occur to create compelling new products or services. This framework also facilitates the collection and organization of information that is possessed across a wide spectrum of participants in these conversations. Each firm should develop its own roadmapping strategy so that it supports the likely interactions that it will encounter over time as it participates in networked innovation. As the context for new product or service innovation changes, the firm may need to modify a previously effective roadmapping strategy to match the new conditions and the potentially different interactions that will be needed.

References

Burgelman, R.: Strategy is Destiny: How Strategy-Making Shapes a Company's Future. The Free Press, New York (2002)

Chesbrough, H.: Open Innovation: The New Imperative for Creating and Profiting from Technology. Harvard Business School Press, Boston (2006)

Choi, T., Dooley, K., Rungtusanatham, M.: Supply networks and complex adaptive systems: Control versus emergence. Journal of Operations Research 19(3), 351–366 (2001)

Gawer, A., Cusumano, M.: Platform Leadership: How Intel, Microsoft and Cisco Drive Industry Innovation. Harvard Business School Press, Boston (2002)

Gulati, R., Nohria, N., Zaheer, A.: Strategic networks. Strategic Management Journal 21(3), 203–215 (2000)

Hult, G., Ketchen, D., Nichols, E.: Organizational learning as a strategic resource management in supply management. Journal of Operations Management 21, 541–556 (2003)

Iansiti, M., Levien, R.: The Keystone Advantage. Harvard Business School Press, Boston (2004)

Kirkpatrick, D.: The Facebook Effect: The Inside Story of the Company that is Connecting the World. Simon & Schuster, New York (2010)

Kostoff, R., Schaller, R.: Science and technology roadmaps. IEEE Transactions on Engineering Management 48(2), 132–143 (2001)

Mintzberg, H., Waters, J.: Of Strategies, Deliberate and Emergent. Strategic Management Journal 6, 257–272 (1985)

Pathak, S., Day, J., Nair, A., Sawaya, W., Kristal, M.: Complexity and adaptivity in supply networks: Building supply network theory using a complex adaptive systems perspective. Decision Sciences 38(4), 547–580 (2007)

Patnaik, D.: The fundamentals of innovation – How IBM, Hallmark and Apple get innovation right – but not for the reasons you might think. Businessweek.com (2010), http://www.businessweek.com/innovate/content/feb2010/id2010 028_823268.html (accessed February 10, 2011)

Petrick, I., Pogrebnyakov, N.: The Challenges in Communities of Creation for Distributed Innovation and Knowledge Sharing. In: Dwivedi, A., Butcher, T. (eds.) Supply Chain Management and Knowledge Management – Integrating Critical Perspectives in Theory and Practice, pp. 3–18. Palgrave Macmillan, Basingstoke (2009)

Phaal, R., Farrukh, C., Probert, R.: Technology roadmapping – A planning framework for evolution and revolution. Technological Forecasting and Social Change 71(1-2), 5–26 (2004)

Stock, J., Boyer, S., Harmon, T.: Research opportunities in supply chain management. Journal of the Academy of Marketing Science 38, 32–41 (2010)

Wu, T.: In the grip of the new monopolists. Wall Street Journal, C3 (November 13-14, 2010)

Wunker, S., Pohle, G.: Built for innovation. Forbes.com (2007), http://www.forbes.com/forbes/2007/1112/137.html (accessed November 12, 2011)

Author

Irene J. Petrick is an internationally recognized expert in strategic roadmapping. She is actively engaged with companies in their innovation and technology strategy activities, including work with twelve Fortune 100 companies, the U.S. military, and a wide variety of small to medium sized enterprises. Petrick has been a professor at The Pennsylvania State University since 2000. In addition to her professorial activities, she has over 25 years of experience in technology planning, management and product development in both the academic and industrial settings. She is author or co-author on more than 120 publications and presentations.

Technology Roadmapping on the Industry Level: Experiences from Canada

Geoff Nimmo

Technology roadmaps in Canada are generally done on a sectoral level. Industry-led and government-supported, these roadmaps are generated by company and academic representatives from the sector and provide a unique private sector perspective concerning future market demands for the sector, and a roadmap as to how and with what technologies companies within the sector will meet these future demands. The timeframe looking outwards is 5-10 years, long enough to get beyond what companies have in their pipeline, but not too long that the projections become fuzzy. The most recent Canadian technology roadmap is slightly different. The Soldier Systems Technology Roadmap, with a government department as the client, is an innovative industry-government collaboration aimed at engaging industry, academia and other research organizations at the front end of Canada's soldier modernization efforts.

1 Introduction

Industry Canada, the Canadian government department responsible for industrial development, launched the technology roadmapping initiative in 1995 as part of its strategic plan to support and promote Canadian innovation. While several different types of roadmaps are existent, Industry Canada chose to focus on sectoral or sub-sectoral level technology roadmaps. The rationale for this decision was to spread the benefits of roadmapping as broadly as possible. Roadmapping on a sectoral level not only includes a wider spectrum of companies and academics, but it also provides a forum where companies in particular can meet and discuss mutual interests.

Industry Canada's technology roadmapping initiative had a single purpose: to strengthen Canadian competitiveness by helping industries to identify and develop the innovative technologies necessary for success.[1] Since 1995, Industry Canada co-sponsored 39 technology roadmaps, with one under development, involving

---

Geoff Nimmo
Roadmapping Consultant,
16 Marlowe Crescent,
Ottawa, Ontario, K1S 1H6
e-mail: geoff.nimmo@gmail.com

[1] Industry Canada discontinued technology roadmapping in June, 2011. However, other departments in the Canadian government are continuing roadmapping.

more than 2700 companies and more than 200 non-industry partners (universities, research institutions and associations). The formation of dynamic partnerships between public and private sector organizations is an important element for success in the evolving marketplace. By stimulating dialogue and collecting valuable information, the technology roadmap process encourages these partnerships and helps establish planning priorities for industry and government.

The collaboration engendered by technology roadmapping extends beyond government and industry. The roadmapping exercise in Canada is a good example of inter-governmental cooperation, with each roadmap involving, on average, at least four governmental departments and agencies. The collaboration between government departments is particularly useful for the implementation of technology roadmaps. Although no single government program exists to assist in turning technology roadmap recommendations into actual projects, there are a number of government programs designed to provide funding across the innovation spectrum. Involving different government departments from the beginning of the technology roadmapping process helps spread knowledge of the roadmap and makes it much easier to interest government programs in the outcome of technology roadmaps.

Within the Government of Canada is an active technology roadmap network, comprised of employees from a cross-section of federal departments who meet regularly to discuss roadmapping and share best practices. The technology roadmap network publicizes the latest developments in technology roadmapping and strives to ensure that the federal government makes the strongest possible contribution to technology roadmapping efforts. Another initiative to spread knowledge and expertise across government and private sectors is the ongoing technology roadmap training program, which seeks to educate government personnel as well as potential facilitators about technology roadmapping.

The following sections outline the technology roadmapping process in Canada in more detail. In the first section, the Canadian technology roadmap process is examined. This includes an explanation of the Canadian technology roadmap model and a description of the three stages of technology roadmaps - development, implementation and evergreening. The next section focuses on the Canadian experience with technology roadmaps, including lessons learned from technology roadmapping. The third section provides advice on determining when and where to develop a technology roadmap. The final section highlights a new form of technology roadmapping recently adopted in Canada, where government is the main client.

2 The Technology Roadmap Process

Traditionally, technology roadmapping in Canada has occurred on sectoral level. The decision to develop sectoral technology roadmaps has much to do with the makeup of the Canadian economy. The vast majority of Canadian companies are small- to medium-sized businesses. While many larger companies conduct their own technology roadmapping or the equivalent because they have the financial and human resources to determine future requirements and solutions, this is not necessarily true of smaller companies who, despite their interest in determining future requirements, generally do not have the time or the resources to undertake such an exercise.

A key objective of this roadmapping approach is to involve the widest range of companies possible from within a particular sector. The most effective roadmaps include a representative cross-section of small-, medium- and large-sized companies. To gain a good range of companies involves a communications strategy, generally done in conjunction with an industry association (if there is an association for that particular sector). As the level of knowledge concerning roadmapping is likely to be relatively low the smaller the company, there will generally need to be presentations and discussions of both the benefits and commitment associated with technology roadmapping (see also the contribution of Phaal, Farrukh and Probert in this book).

Canada's technology roadmapping strategy creates a collaborative environment that gives smaller companies the opportunity to identify future requirements and 'rub shoulders' with larger companies. The networking opportunities and partnerships that are established during the development phase are key benefits of the technology roadmap process. For example, larger companies within the aerospace sector are generally integrators, and look for companies to source their products. Smaller companies with good products can take advantage of the networking opportunities the technology roadmap process offers and gain visibility with larger companies. Up and down the supply chain, future business arrangements are discussed and implemented. For Industry Canada, whose key clients are Canadian industries, this is a desired outcome. All technology roadmap evaluations conducted by Industry Canada highlight this particular benefit.

Workshops are a key aspect of technology roadmapping. Whenever possible, the workshops involve participants being seated around 8-10 person roundtables. During the one or two day workshops, conversations and friendships ensue and knowledge of other companies and their capabilities are greatly increased. A common occurrence with completed technology roadmaps is that companies have far more knowledge about other companies in their sector.

The commitment being asked of companies for a technology roadmap are both fiscal and human resource-related. Technology roadmaps generally take approximately a year to complete, and involve about four workshops. While the overall costs of the technology roadmap are covered by government, industry participants pay in kind through their attendance and substantive inputs during the workshops. Particularly for smaller companies, attendance at all workshops is not an insubstantial cost.

One might question why larger companies would want to be involved in roadmap development in the Canadian context, particularly if they already develop technology roadmaps internally. Canada has never experienced problems with larger companies not wanting to be involved in technology roadmaps. Roadmaps have been developed in some sectors where there are very few large companies, and many smaller ones. Even in these extreme cases, the larger companies did not hesitate to join the collaborative process. Although specific reasons for participating were not explored in detail, the driving factors could be the government's involvement in the process, the desire to meet possible suppliers, and the company's desire to be perceived as a willing participant. Another possible factor is that smaller companies tend to be more flexible and agile in their thinking, and this could be of interest to larger companies.

Regardless of an individual company's motives for participation, it is essential to have larger players involved in the technology roadmapping process. While they cannot claim to possess all the knowledge and wisdom of the sector, larger companies do provide the weight and name recognition that can attract other companies and pique government interest.

2.1 The Canadian Technology Roadmapping Model

Figure 1 provides a graphical representation of how Industry Canada traditionally develops technology roadmaps. The left side of the diagram, Who? illustrates how government, academia and research organizations feed into the industry suppliers, manufacturers and end-users who are the key industry drivers in this process. The middle section, the What/How shows the three phases of the roadmapping process- the development of the technology roadmap; the implementation of the developed technology roadmap; and the evergreening of the technology roadmap. The right side of the diagram, Why, illustrates the anticipated benefits for industry, academia and government from the roadmapping process.

Key to roadmapping in Canada is who does what. While government acts as a catalyst and process expert, it is industry that leads the technology roadmapping process and makes all major decisions. The roadmap is industry's roadmap, which means that the substantive knowledge and recommendations contained within are industry's inputs. What government wants is a roadmap that is a clear voice from industry - one that includes ideas and recommendations on what the sector needs to do to be more competitive in the future.

It is made clear to industry at the beginning of the process that the technology roadmap is as the title suggests- a document that will provide clear direction as to the technologies that will be critical for companies in the sector to meet future

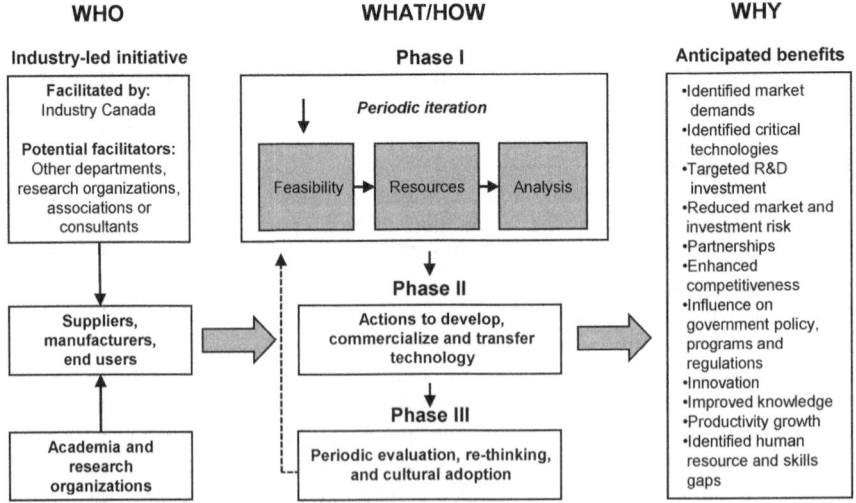

Fig. 1 The Canadian model of technology roadmapping

market demands. It is not intended to be a policy document, and developing a technology roadmap to achieve policy goals will severely lessen its value. While government policy may be influenced by the recommendations on future technologies and skill requirements, policy development is a by-product and not the primary purpose.

2.2 The Three Phases of the Technology Roadmap

Although the planning horizon for a technology roadmap will differ from one sector to another, depending on the speed of technology development, the steps for technology roadmapping remain the same. The technology roadmapping process consists of three separate phases: technology roadmap development, implementation and evergreening.

Phase 1 - Technology roadmap development: Phase 1 begins with determining the feasibility of developing a technology roadmap within a particular sector. An overview study of the sector is first undertaken. The study provides current information on Canadian company capabilities, international market research, the state of technological innovation, emerging market trends, sector-specific productivity statistics as well as human resources and skills/training requirements. In order to project forward, it is necessary to know where the sector is at the moment- that is the intent of the overview study.

The next step includes identifying and gaining the commitment from industry leaders who are well positioned to lead the process. Collaborating with industry, the task is then to bring together the right minds- industry experts, academics, technology researchers, analysts, economists, and government specialists- who will bring insights to the table. The commitment by industry to the roadmapping process is fundamental. Is the industry genuinely determined to develop the technology roadmap? Does commitment come from a representative cross-section of companies within the sector? Measures of industry interest could include the amount and level of resources and time committed by industry members and associations at this early stage - for example, individual companies stating their anticipated level of commitment in person-days. Another indicator is the number and relative "importance" of industry members involved at this stage.

Technology roadmaps that have been least successful in Canada are those where industry interest diminished as the roadmap proceeded, leaving the final writing of the roadmap in the hands of government. This is not the intent of technology roadmapping, and can result in significant problems. The driving force and substantive knowledge for the roadmaps must come from industry. This is not to say that government does not fulfill valuable roles in the roadmapping process- government representatives help organize and facilitate the Workshops, report progress and assist in agenda-setting for next steps. They also, when requested, provide policy, regulatory and other industry-specific information that could influence the direction of the roadmap. Finally, government participation ensures that important information is brought back to policy makers and to R&D funding organizations about the progress of technology roadmaps, and how government can encourage greater innovation in Canadian industry.

The next step is the identification of an industry champion and his or her acceptance of the position. Industry champions are a critical part of the process, as they become the public face of the roadmaps and play a major role in its development and publicizing of results. Industry champions should have the stature and reputation in the industry to quickly resolve any issues - at times by making a single phone call - that might arise. These individuals are appointed Chair of the Steering Committee, which generally consists of 8 to 12 members, approximately two-thirds of whom represent companies from the sector.

Creating a technology roadmap is a collaborative, iterative process. Although the planning horizon differs from sector to sector, the steps for technology roadmapping remain the same. Technology roadmaps are developed through a series of workshops (generally four) that take place over the course of a year. The first, a Visioning workshop, begins with recognizing and identifying that a problem exists, and understanding that it can be resolved through technology roadmapping. Working collaboratively, participating firms identify the markets and needs that will fuel the industry's growth in the next three to ten years. This vision reflects the dynamic impact of market, technology, skills and regulatory drivers, and articulates key industry goals. This is a demand-side forecast, based on what companies and academics think that customers and markets will demand.

The level of representation at this first workshop is important. Senior officials from participating companies should be present at this initial meeting. With the development process lasting approximately one year, the involvement and interest of senior management at the outset sends a clear message about the technology roadmap's value to the company. While participation of more technical personnel is preferable for subsequent workshops, the raised level of participation at the first workshop increases the likelihood of continued participation. Senior personnel who attend the first workshop can be kept informed of all future meetings and technology roadmap progress by other, less-senior personnel involved in the process. The size of the company can also affect company representation- smaller companies having limited resources can impact their involvement.

The participation of company executives - presidents and vice-presidents - in the Visioning workshop also makes sense from a content perspective. To effectively guide their companies, senior-level executives should understand and be able to communicate their vision of future market demands. Intellectual property is not an issue in the Visioning workshop- this is a forward projection that does not involve discussion of company products or technologies.

All workshops, including the Visioning workshop, are guided by a lead facilitator. This individual, generally a consultant with strong knowledge of and experience in technology roadmapping, plays a key role in technology roadmap development. There are a number of experienced facilitators in Canada. These individuals use their experience to help develop workshop agendas and facilitate the sessions to ensure that all are heard and that consensus is formed. At the conclusion of each workshop, the facilitator records the results and evaluates what worked well and what should be adjusted for the next session.

What is the profile of a facilitator? The facilitator should be an expert in process but not necessarily in content - although for certain technical sectors, there is a need for familiarity with the subject matter. The individual must understand the technology roadmapping methodology and be able to bring out the best

Technology Roadmapping on the Industry Level: Experiences from Canada   53

possible contribution from all participants. To successfully fulfill his or her role, the facilitator needs to remain impartial to the issues under discussion, keeping a distance from the content. The facilitator remains involved until the roadmap has been completed and approved.

In addition to knowing the process, facilitators must be skilled communicators and listeners. Generally, workshop participants will not be familiar with technology roadmapping, and will not be accustomed to working and collaborating with other companies - some of whom might be their direct competitors. A facilitator must be able to explain workshop goals, and to impart the understanding that this is a process from which everyone can gain.

Typically, the second workshop takes place two to three months following the Visioning workshop. The objective of this second session is to determine the characteristics of the products (supply side) to be developed by the industry in order for sector firms to compete successfully and meet future market demands. Given that this step could deal with proprietary information (as companies need to examine their next-generation products to determine need), participating firms need only disclose information they feel comfortable sharing. When firms are reluctant to share information, the government is considered an honest broker, and information can be shared with government but not competitor companies.

The objective of the third workshop is to determine key or critical technologies that must be in place to competitively design, manufacture and support the proposed future products. Technologies should be described in sufficient detail so that sector firms can evaluate their current capabilities, identify their own technology gaps, and determine how to bridge those gaps. The technology descriptions do not deal with proprietary products or information and therefore can be shared. In effect, the technologies become the technology roadmap- the recommended actions to ensure that companies in the sector are prepared to meet the future market demands.

Prior to the fourth workshop, the results of the previous workshops are pulled together, along with the initial sector overview, to produce a sectoral technology roadmap. The recommendations in the roadmap generally concern pre-competitive enabling research that will allow firms to innovate and generate next generation products. The fourth workshop is a testing exercise- the draft roadmap is presented to participants and is open for discussion. Following the meeting, the document is also sent to as many companies within the sector as possible- the technology roadmap is industry's document, and it must be a strategy that applies to all companies within the sector. Phase 1 is complete upon the development of a formal technology roadmap - a document that reflects the commitment, decisions and directions of industry.

Phase 2 - Implementation: Once the technology roadmap is completed, it becomes the basis for cooperative research, development and deployment activities that focus on new technologies and skills. The roadmapping process encourages participants to align their research and development efforts with the high-priority needs identified in the roadmap. This approach maximizes investments by ensuring the most strategic allocation of scarce resources while accelerating the research and development process.

For those companies or collaborative ventures that are looking for assistance in moving ahead with technology development, there are funding possibilities. While there is no specific funding program in place for technology roadmaps, there are government programs across the Canadian government to assist with innovation and technology development. A difficulty is that a number of the programs are not well known, particularly by smaller companies.

The technology roadmapping implementation strategy employed by Industry Canada is designed to introduce the government innovation programs to industry members. A compendium of government programs and services for the advancement of technology roadmaps was developed listing 350 federal and provincial programs, prioritized by their relevance to technology development, that could assist in the implementation of technology roadmaps. Now, when a roadmap is complete, roadmap participants can attempt to align their technology requirements with the goals/requirements of innovation funding programs.

Phase 3 - Evergreening: The third phase of the TRM process consists of periodically reviewing and updating the technology roadmap. Roadmaps will always be "works in progress", and must be revisited - as market demands change, as Canadian industry expands into new market niches, as regulatory changes shift the technology focus, and as new technologies mature. Changes in technology will likely necessitate updating and altering the skills requirements for the sector. By this phase, industry partners have adopted technology roadmapping as a standard business planning tool. By keeping their roadmaps "ever green", companies remain focused on future markets and the technological innovations that will help them compete and win. The evergreening process does not demand a new technology roadmap. Generally the updates can be done relatively quickly over one or perhaps two Workshops.

3 Determining When and Where to Develop a Technology Roadmap

There are always questions about how to determine if a sector or sub-sector is ready for a technology roadmap. Are there certain elements to look for in determining the readiness of a sector? How can you judge between competing sectors?

Some key conditions that can help determine a sector's need for a technology roadmap would include:

- market demands are changing dramatically;
- the industry has reached a strategic juncture regarding entering new markets, seeking out new technologies or acquiring new skills;
- companies within the industry are losing or failing to increase market share as new markets emerge or the latest competitive threats arise;
- companies have a vision of their place in future markets but no strategy to turn vision into reality; and
- companies are uncertain about which technologies and applications future markets will demand - and when they will be needed.

While the need for a technology roadmap may be apparent, there are still critical issues in order for a sector to be selected for a technology roadmap.

For industry:

- The sector presents a clear need and shows significant interest in a roadmap;
- the sector is facing critical challenges;
- the potential for research and development is significant;
- the commitment across the sector is real and substantial, and includes all size of company; and
- the sector commits to active involvement in the implementation phase.

For government:

- The project scope is aligned with governmental priorities;
- there are sufficient short- and long-term resources committed to the project both internally and externally; and
- commitment to the roadmap exists at all levels in government and industry.

4 Canadian Experiences with Technology Roadmaps

As Figure 2 shows, roadmaps have covered vast areas of the Canadian industrial spectrum. From aerospace and oil sands to wireless technology, these examples prove that successful technology roadmaps can be developed when commitment and desire exist on the part of Canadian industry. With roadmapping beginning in 1995, the first pilot roadmaps were completed in 1996.

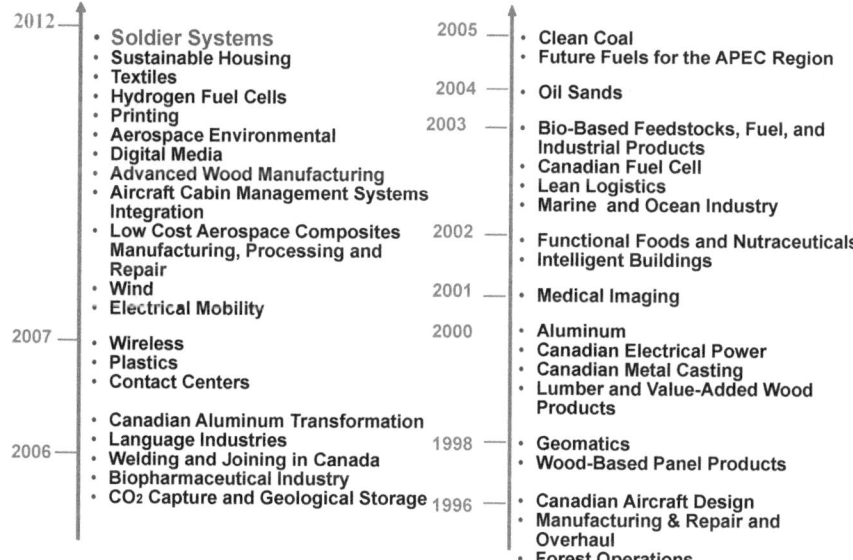

Fig. 2 Status of Canadian technology roadmaps

Technology roadmapping should benefit all participants. For companies, the roadmap is a strategic planning tool that identifies gaps between current technological capabilities and future requirements. It is also instrumental in helping firms make informed investment decisions to close gaps between current capabilities and future needs. For research organizations and educational institutions, the roadmap provides guidance for structuring future programs. For government, it provides a strategic direction for industrial development activities.

Although termed "technology" roadmaps, in some cases the technologies were not always the key focus. For example, in 2003 four technology roadmaps were requested to be developed in areas that could lead to significant environmental improvements - mitigating emissions through the development and introduction of environmental technologies. One of the areas chosen was Fuel Cells. Given the significant size of the fuel cell industry in Canada in 2003, it was decided that a roadmap should be developed to determine the best way to move to commercialize fuel cell technologies. The traditional roadmapping process was followed: identifying where the fuel cell sector needed to be in five to 10 years; determining the gaps and challenges to move forward to commercialization; and determining a strategy to overcome the obstacles. The unique feature was that a commercialization path took the place of a technology development path.

Technology roadmapping can also lead to useful collaborations with other future-oriented planning processes. In Canada, Human Resources and Skills Development Canada (HRSDC) is responsible for skills development. One of the tools the department employs are Sector Councils, which promote a sectoral, future-oriented, industry-driven approach to future workplace skills needs and development. Recognizing the similarities between the work of sector councils and technology roadmaps, the departments have collaborated on the development of five roadmaps (Wireless; Contact Centres; Printing; Digital Media; and Advanced Wood Manufacturing). With skills and technology being two sides of the same coin, the collaboration between the programs demonstrates that the process itself can be mutually reinforcing with multiple innovation/productivity objectives.

On an international level, Canadian knowledge of roadmapping and technology foresight led to the development of the Future Fuels Technology Roadmap. This roadmap included over half of the APEC (Asia-Pacific Economic Cooperation) countries in a collaborative effort to determine a secure and sustainable energy supply for the APEC region for the foreseeable future (out to 2030). A staple of technology foresight, scenario creation, was used for understanding the role of emerging energy fuel technologies in energy futures. Once these technologies were identified, technology roadmapping was employed to identify critical steps in the development of these technologies.

While no two roadmaps are exactly alike, one constant factor is the level and extent of collaboration - among participating companies; between industry and government in the preparation and implementation of the roadmap; among government departments; and between countries. The benefits of working together seem to be clear to participants, no matter the grouping.

4.1 Specific Examples of Technologies/Collaborations That Have Emerged through Technology Roadmaps

There are numerous examples of concrete results from Canadian technology roadmaps. Two roadmaps, one from the manufacturing and one from the service sectors have been used to illustrate the tangible benefits that can result from technology roadmapping.

The *Canadian Aircraft Design, Manufacturing and Repair & Overhaul Technology Roadmap*, completed in 1996, was developed to identify the critical enabling technologies the sector required to design, build, and maintain aircraft, aircraft systems and components to meet customer requirements in the period 2001-05. Technologies were selected on the basis of their potential contribution to marketplace competitiveness and their strategic applicability across the industry sector. Completed in 1996, the roadmap delivered the following results:

- *Strategic planning* - the Canadian National Research Council's Institute for Aerospace Research used the roadmap to structure its 1999–2004 Strategic Plan;
- *Partnerships* - the Aerospace Industries Association of Canada established the Office of Collaborative Technology Development to promote collaborative technology development; and
- *Collaborative R&D* - a program was introduced to support collaborative R&D projects.

The *Lean Logistics Technology Roadmap* was completed in 2003. The intent of the roadmap was to improve the productivity level and supply chain technology adoption of Canadian SME (small- and medium-sized enterprises) suppliers to large North American customers. The completed roadmap had the following recommendations and results.

- *Collaboration* - the government department responsible for skills development, Human Resources and Skills Development, and industry committed more than $1 million to fund a national skills study that established the foundation for a Logistics Sector Council;
- *Policy* - guidelines were developed to assist Canadian SMEs with outstanding border (Canada-United States) compliance issues;
- *Networking* - a supply chain technology pilot project was introduced between a national grocery store chain and its SME suppliers (with financial support from Transport Canada); and
- *Horizontal impact* - an annual report on supply chain e-technology adoption across all industry was initiated.

4.2 Lessons Learned from Technology Roadmapping

1. *All sectors are unique.* Every industry sector or sub-sector displays different characteristics that will necessitate slight deviations in process. While all roadmaps adhere to an underlying framework (demand driven; workshops), sectors have differing structures and requirements which lead to variations in needs and process design.
2. *The value of the industry champion.* While it is not always obvious where to find the ideal industry champion, there is no doubt of the valuable role the champion plays. These individuals can be selected from a small company, or from an industry giant. Regardless of where an industry champion is found, that individual must be a well respected figure within the industry and be on a first-name basis with large and small companies within the sector. Industry champions are also critical to the process after the roadmap is complete. At that point, the technology roadmap becomes a communication document that industry - usually through the champion - can use to approach and attract the interest of government. With the support of the roadmap, the champion can demonstrate how industry has come together in a collaborative venture to identify technologies that are critical to sustaining future competitive success.
3. *The importance of industry commitment.* Industry leadership and inputs are critical to good technology roadmaps. While understanding how busy industry is, it is important to keep industry actively participating throughout the roadmapping process. Technology roadmaps that last longer than a year have a difficult time retaining industry participants.
4. *Technology roadmaps are not only about technology development.* Key to improving the roadmapping process is understanding what has worked and not worked in earlier roadmaps. In the eleven TRM evaluations undertaken by Industry Canada, a constant theme is the value that TRM participants place on the Development phase workshops. When possible, the workshops feature 8-10 person roundtables. The roundtables lead to many discussions, which in turn lead to greater knowledge of other companies in the sector, and possibilities of acquisition and supply. This human contact has real benefits. Up and down the supply chain, companies can gain from the process. For small- and medium-sized companies, roadmapping provides a means of determining what larger companies are seeking; for large companies that operate as integrators, roadmapping provides a new source of possible suppliers. Greater knowledge leads to increased opportunities.
5. *The importance of skills.* In determining what is required to meet future market demands and become increasingly competitive, the issue of future skills needs arises in all roadmaps. Skills are a natural complement to required technologies - in essence, two sides of the same coin.
6. *Difficulties of implementation.* The most challenging aspect of technology roadmapping is implementation. Responsibility for implementation activities rests with the roadmap participants. The roadmap document should provide enough information to make technology selection and investment decisions.

Technology Roadmapping on the Industry Level: Experiences from Canada 59

Collaboration on technology development is preferred, but if a critical mass of companies is not prepared to collaborate on research and development and share the rights to the resulting new technologies, individual companies may undertake their own research and development projects. Regardless, roadmap participants are made aware at the beginning of the roadmap process that there is no funding program attached to the roadmap, and that they should be considering how to undertake implementation throughout the roadmap process.

5 An Alternative Form of Technology Roadmapping in Canada

Based on the experience gained through developing technology roadmaps, an alternative form of technology roadmapping was developed in Canada where government becomes the client. The following description of the development and implementation of the Soldier Systems technology roadmap illustrates how traditional roadmapping methodology can be adapted in novel and exciting ways.

5.1 Government as the Client – The Soldier Systems Technology Roadmap

The traditional form of technology roadmapping will continue to exist in Canada. However, over the past four years, a technology roadmap has been developed that is different in scope and execution from previous roadmaps. Created in conjunction with the Department of National Defence (DND), this roadmap focuses on helping to equip the Canadian Soldier of the Future.

The *Soldier Systems Technology Roadmap* is an innovative industry-government collaboration aimed at engaging industry, academia and other research organizations at the front end of Canada's soldier modernization efforts. The sharing of knowledge and exploring what is possible in soldier systems technologies is designed to help provide Canada's Soldiers of the Future with the best and most practical equipment.

The two-way communication with industry helps to ensure a better match between the Canadian Forces' modernization goals and what industry can deliver over the next 5-10-15 years. In certain ways, the development of the Soldier Systems Technology Roadmap is similar to other roadmaps. It is being developed through a series of workshops, including an initial visioning exercise, followed by more technical and focused workshops, and concluding with a document that captures all the workshop results. The differences from the more traditional roadmaps, however, are significant.

Traditional technology roadmaps are industry-driven, but in the case of the Soldier Systems Technology Roadmap, the Department of National Defence becomes the ultimate client. The roadmapping exercise is driven by what the Department of National Defence has determined for their future requirements. Having previously undertaken a large Mind-Mapping exercise, the Department of National Defence identified approximately 900 technologies needed to support future requirements across a number of strategic areas. While the technologies had been identified, there was no easy way to prioritize them or identify the technology-readiness level. Knowledge was also lacking concerning which industries have understanding and capability in these areas.

The Soldier Systems Technology Roadmap allows the Department of National Defence to gain knowledge and expertise from industry, and to develop insights into obtainable technologies and their maturity levels to guide the development of realistic and achievable Canadian Forces acquisitions requirements. The roadmap becomes a platform for the Department of National Defence to better align technology development with its needs by providing industry and other defence stakeholders with knowledge about the Department of National Defence's priorities for the future. Overall, it provides a platform to better synchronize industry and government technology planning to leverage limited research and development funding and better meet Canadian Forces needs.

It is important to understand that the Soldier Systems Technology Roadmap is not about procurement. Procurement and acquisition will occur in the future in the usual manner. The roadmap is about the exchange of information related to the Department of National Defence's future requirements, and the research and development that will take place to position companies or academics for the future procurement opportunities. This exchange of information benefits all sides by providing intelligence about the Department of National Defence requirements, and in turn, receiving industry advice concerning these requirements.

Implementation of the Soldier Systems Technology Roadmap will also involve government-industry collaboration. While the responsibility for implementation rests with the private sector, the government will provide the infrastructure to assist with implementation. Through the Workshops, the Department of National Defence informed industry and academia about their future needs. Companies and academics, individually or collaboratively, will develop proposals to undertake the R&D necessary to put themselves in a position to provide what the Department of National Defence is requiring in the future. In turn, government has created an interdepartmental Management Office to receive proposals and to respond quickly to the originators of the proposals as to the alignment of the proposal to the Department of National Defence's future needs.

5.2 Soldier Systems Technology Roadmap Enablers

The Soldier Systems Technology Roadmap was a novel experience for Canada. Not only were there no previous technology roadmaps where government was the client, but there were no other technology roadmaps developed on such a vast scale. For this technology roadmap, there was a need for roadmap enablers to ensure that the roadmap was accessible to all. The enablers included the Workshops (Visioning and Technical Workshops) and a software collaboration tool called the Innovation, Collaboration and exchange environment (ICee). Each of the tools had a unique role to play in the Development Phase of the Soldier Systems Technology Roadmap.

Visioning and technical workshops
- Face to face meetings with industry/academia/gov. researchers

Collaboration Tool (ICee)
- Database linked with WIKI for sharing/collaboration on technologies
- Developed by IC to capture DND needs & industry capabilities
- First public and password controlled Wiki in Canada
- Vetted and approved by Government of Canada (Dept. of Justice)

Unique Technology Management/Roadmapping Software
- Visual depiction of links and synergies between all roadmap items

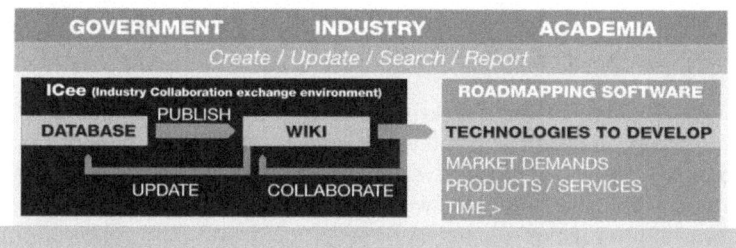

Fig. 3 Technology roadmapping software solution

5.3 Visioning and Technical Workshops

The Soldier Systems Technology Roadmap is the largest and most complex technology roadmap that has been developed in Canada. With the active involvement of five federal government departments, industry and academia, the subject areas for the roadmap include Power/Energy; Lethal and Non-Lethal Weapons; C4I (Command; Control; Communications; Computers and (military) Intelligence); Sensors; Personal and Protective Equipment; and Human and Systems Integration.

For each of the subject areas, there was a 2-day Technical Workshop. Split into four sections, the workshops were carefully structured to ensure that there would be clear deliverables at workshop's conclusion. Each Workshop began with a series of presentations by Subject Matter Experts from the Department of National Defence explaining the future requirements for the particular area. With the registrants carefully arranged around 10-person roundtables (participants were arranged to ensure a good cross section of industry, academia and government at each roundtable), responses to the Department of National Defence's future requirements were elicited. The second part of the Workshop concerned the gaps/challenges between the Department of National Defence's future requirements and what was the situation at present. The goal was to determine as clearly as possible the gap that needed to be bridged. There was again a series of presentations by the Department of National Defence, and responses by the roundtables. The last two parts of the Workshops concerned what were the technologies that could bridge the gaps, and who was presently working on the most important of these technology areas.

The Technical Workshops were very successful. Over 1200 industry, government and academic personnel participated in the workshops. Collaboratively, they identified 20 Theme areas across the subject areas. For each Theme area, there was a clear Theme objective (the specific change that has to occur within this theme area in order to contribute to the technical domain's overall system goal and to address the identified gaps); Technical Challenges and Requirements (the technology problem that needs to be overcome to achieve the objective, including a description of the technical requirement); Enabling Technologies (the specific technology areas that need to be adjusted or improved to overcome the challenges-some of these are traditional technology areas and some are emerging technologies); and Research and Development focus areas (the specific areas of research and development that need to be pursued within the enabling technologies. which also includes a description of known research areas in the area).

The results from all the Workshops have been compiled into a 404 page Capstone Report and Action Plan that has recently been made publicly available. The report provides a comprehensive analysis of future requirements and where it is hoped industry and academia will focus their efforts to meet those future requirements. Information on how to access the report, and other information about the Soldier Systems Technology Roadmap implementation, can be found on the Soldier Systems website (www.materiel.forces.gc.ca/en/sstrm.page).

5.4 The Innovation, Collaboration and exchange environment (ICee)

The Technical Workshops were a critical element of the Development Phase of the Soldier Systems Technology Roadmap. Because implementation will be stretching into the future, it was determined that software was needed to facilitate and continue the collaborative aspects of the roadmapping process. The ICee was developed to create an effective interface between industry, academia and government. Launched October 1, 2009, the ICee represents the Government of Canada's first outward-facing Wiki. Developed in-house at Industry Canada (IC) and created through strong departmental cooperation, this innovative web tool is already attracting attention as a model for other applications.

The ICee consists of two main components:

1. The ICee database: a password-protected database used for collecting key information on roadmap items (e.g. information about technology needs and capabilities); and
2. The ICee Wiki: a web-based forum for participants to comment and collaborate on technology roadmap items (e.g. a capability; technology project; product or service). This information is integrated into the roadmapping process.

Over a secure and password-protected environment, the ICee allows users to search and share technology information that can help Canada's soldier modernization effort. The ICee is a web-based knowledge transfer tool that enables industry, academic and government participants to exchange information on emerging technologies and play an active role in developing the Soldier Systems Technology Roadmap. Another function of the ICee is that it levels the playing field for all of Canadian industry- while not all companies had the opportunity to attend the workshops, all have the opportunity to access the ICee. None of the material from the Soldier Systems Technology Roadmap is classified or confidential- all of the information from all of the workshops is included on the ICee.

With well over 700 registered users, the ICee promotes collaboration, and extends the reach of the Soldier Systems Technology Roadmap by facilitating online participation and interaction- especially among experts from small-and -medium-sized enterprises and research organizations located in distant or international locations. The ICee complements the normal formal exchange mechanisms of a technology roadmap (e.g. workshops; meetings; briefings) and will be an ongoing information/collaboration exchange tool now that the development phase of the roadmap has been completed.

6 Conclusion

Companies face tremendous business challenges today. Virtually all products, services and operations depend on rapidly changing technologies. Products are becoming more complex and consumers more demanding. Product life cycles are shortening and product time-to-market is shrinking. High-calibre, innovative competition abounds.

It's no secret that the companies with the greatest productivity and largest market share know how to forecast, analyze and plan. To remain competitive and ensure their long-term success, companies must focus on their future markets and apply a well-researched technology development strategy. This is where technology roadmapping comes in.

The technology roadmapping concept is at heart a consultative process. Its primary objective is helping industry, its supply-chain, academic and research groups, as well as governments come together to identify and prioritize the technologies needed to support strategic R&D, marketing and investment decisions. For sectoral technology roadmaps, these are technologies that will be critical to an industry five to ten years into the future.

The approach of the Soldier Systems Technology Roadmap is simple- to understand how the technologies of today, and the technological possibilities of tomorrow, can contribute to the individual soldier's future capabilities. By focusing on the technology needs of the soldier as the center of a complex and integrated system, the Soldier Systems concept incorporates anything related to the life and work of a ground force combatant. Roadmapping is central to this approach by providing a proven approach for allowing industry, academia and government to work together to identify technology needs and a strategy for addressing those needs.

Through the first 17 years of Canada's technology roadmapping initiative, lessons have been learned and alternative approaches adopted. The alternatives do not detract from the value of technology roadmapping; rather they show the adaptability and strength of this strategic planning process. The value of roadmapping as a tool to bring together industry, academia and government to collaboratively address challenges and opportunities has been clearly established by the ongoing and enthusiastic collaboration of the parties. This is, and will remain, the central feature of all forms of roadmapping in Canada.

References

Garcia, M.L., Bray, O.: Fundamentals of Technology Roadmapping. Sandia National Laboratories, Alburquerque (1997)

Government of Australia, Technology Planning for Business Competitiveness - A Guide to Developing Technology Roadmaps. E.I. Section, Department of Industry, Science and Resources (2001), http://www.technologyforge.net/enma/6020/6020Lectures/TechnologyRoadmapping/ENMA291TRReferences/TechnologyRoadmapping.pdf (accessed December 15, 2011)

Government of Canada, The Future Security Environment 2008-2030. Department of National Defence (2009), http://www.cfd-cdf.forces.gc.ca/documents/CFD%20FSE/Signed_Eng_FSE_10Jul09_eng.pdf (accessed December 15, 2011)

Industry Canada, Technology Roadmapping in Canada: A Development Guide. Industry Canada (2007), http://www.ic.gc.ca/eic/site/trm-crt.nsf/vwapj/development-developpement_eng.pdf/$file/development-developpement_eng.pdf (accessed December 15, 2011)

Author

Geoff Nimmo graduated from the University of Victoria with a BA in International Relations in 1979, and from Carleton University in Ottawa, Canada, with a Masters in International Relations in 1981. For the past 11 years, Geoff has been Manager of the Technology Roadmap Secretariat at Industry Canada in the Canadian Government. In that capacity, he helped develop and implement technology roadmaps in many sectors of the Canadian economy. Most recently, the focus of the Technology Roadmap Secretariat has been the Soldier Systems Technology Roadmap, a collaborative venture between the Department of National Defence, Canadian industry and the academic community. Mr. Nimmo has now left the Canadian Public Service and is working on technology roadmaps as a consultant.

Roadmapping as a Responsive Mode to Government Policy: A Goal-Orientated Approach to Realising a Vision

Clive I.V. Kerr, Robert Phaal, and David R. Probert

Government policy documents such as white/command papers embody a country's future vision for its public services (e.g. defence, energy, health, transport). It is then the task of specific government departments/agencies to realise such visions through their strategic planning activities. To aid departments in their crafting of responses to newly issued policies, the use of roadmapping is proposed as a visual tool to facilitate the elicitation process of determining the most appropriate course of action. To demonstrate this goal-orientated approach, a case study based on the Australian Government's Defence White Paper and the Royal Australian Navy's fleet plan will be presented. The developed roadmap employed a new form of architecture, which consisted of a composite structure, in order to provide a logical decomposition of the government's future vision against the major projects to be conducted as the route to policy implementation. The process to populate the roadmap will be outlined together with a description of the roadmap canvas with its associated visual objects. It is hoped that the roadmap presented in this chapter will act as a graphical datum/prototype for utilising roadmapping in a responsive mode to policy directives.

1 Introduction

A government's vision for its public services are often outlined in white/command papers. This form of policy document provides a future end-state at a national level and it is then the task of specific government departments/agencies to implement these policies in order to realise the stated vision. In response to a policy vision, and thus shifting the emphasis to its actual implementation, a critical activity is the strategic planning that takes place between the different stakeholder

---

Clive I.V. Kerr · Robert Phaal · David R. Probert
Centre for Technology Management,
Institute for Manufacturing,
Department of Engineering,
University of Cambridge,
17 Charles Babbage Road,
Cambridge, CB3 0FS,
United Kingdom
e-mail: civk2@cam.ac.uk,
         {rp108,drp}@eng.cam.ac.uk

groups. Given that strategic planning is one of the most demanding tasks faced by managers, the work of Eppler and Platts (2009) on *'visual strategising'* advocates the utilisation of graphical devices. Such management visuals provide a cognitive aid that supports communication and decision-making. This chapter will propose how roadmapping, as a visual-based management support tool, can be applied by government to translate policy documents into strategic plans that satisfy national objectives.

To realise policy visions, a goal-orientated or *'normative'* approach should be adopted. One such approach is outlined in this chapter which provides a logic-based, top-down decomposition by bringing together concept mapping, backcasting and path dependency theory. This normative roadmapping process will be described along with its corresponding roadmap template that needs to be populated. To demonstrate its application, an illustrative case study based on real-world data along the time horizon of 2010-2030 from the Royal Australian Navy and the Australian Government's latest Defence White Paper (DoD 2009a) is presented. The policy paper was launched by the Australian Prime Minister Kevin Rudd on the 2nd May 2009 (DoD 2009b). It provided a rich case because "defence planning is, by its very nature, a complex and long-term business" (DoD 2009a). Further, it is a very relevant case study with ample opportunities for learning because the defence domain is one area of public policy planning where decisions taken in one decade have the potential to affect, for good or ill, a government's freedom of action for decades to come (DoD 2009a).

2 Applying Roadmapping

In Eppler and Platts' (2009) conceptual framework for visual strategising, which recognises four distinct visual practices/genres, roadmapping is classified as a technique under sequencing methods within the planning phase. Visual diagrammatic-based representations, such as roadmaps, are "ubiquitous as aids to human reasoning" (Chandrasekaran *et al.* 2004). "If the definition of a roadmap is generalised to being a visualisation of strategy or strategy elements, then the use of roadmaps can be extended to support any decision process" (Whalen 2007). Thus, from a management tool perspective, roadmapping is indeed a powerful technique for supporting planning – especially when exploring the linkages between resources, organisational objectives and the environment (Phaal *et al.* 2004a). It allows for an identification of gaps, prioritisation of issues and action plan development (Gindy *et al.* 2006). Roadmapping also provides a mechanism to link the strategic level of decision-making down through to the operational levels (Petrick and Provance 2005). However, roadmaps are as much a communications tool as a planning tool (Ma *et al.* 2006). This feature of roadmapping is very evident in the 'strategic lens' type of roadmap (Phaal and Muller 2009).

Taking a social sciences perspective, Kerr *et al.*'s (2009) conceptual framework postulated that roadmapping and roadmaps provide a mechanism/vehicle to cogitate, articulate and communicate. Under this psychosocial interpretative lens, the *'communicate'* verb represents roadmapping as an activity for participants to actively converse and a roadmap as a vehicle to convey the results of the activity and connect with the array of stakeholders. It is the visual patterns in the roadmap, through text/symbols/colour etc., that thus give meaning. Kerr *et al.* (2009) were the first to consider a roadmap as a boundary object since "it is a communication device that not only helps to share information to other parties for strategic dialogue but to more importantly mobilise action". Star and Griesemer (1989) developed the concept of boundary objects for artefacts that exist at "junctures where varied social worlds meet in an arena of mutual concern" (Clarke 2005) – as is the case of roadmaps for communication and co-ordination in strategic planning activities. Borrowing from the domain of informal logic and argumentation theory for the basic principles of constructing sound arguments, the meaning embodied in a roadmap for communication purposes can be seen as an action-seeking type of dialogue. Mann (1988) defines this as a situation where the goal of one party is to steer another party and direct them to carry out a specific course of action. In this case, it is the Department of Defence directing the Royal Australian Navy to implement the Government's defence policy. Under this type of dialogue the aim is to co-ordinate goals and actions (Walton 2008), i.e. how to implement the vision of national defence for *'Force 2030'*. Thus, the objective of the actual dialogue is to decide and depict the best available course of action (Walton 2008); hence, the rationale for proposing roadmaps as a visual tool to facilitate the translation of policy visioning documents into strategic implementation plans.

To demonstrate the application of a roadmap as a tool for depicting and then communicating strategic plans in response to issued government policy, a reference case has been developed utilising real data on the national defence of Australia. To understand the context and challenges faced in the implementation of defence policy, consider Figure 1 which gives a visual depiction of the problem statement. At the macro level there are three principal stakeholder groups, i.e. the government, the military and industry. The government is represented by the Department of Defence; the military represents the end-user, e.g. the Navy; and, industry are the product-service-technology providers. To realise the vision embodied by the Australian Defence White Paper (DoD 2009a), there is the challenge of aligning current and potential future military capabilities to the specified future environment. This requires a balance/compromise being made between the operational needs of the warfighter, the budgetary constraints imposed by central government upon the Department of Defence and the availability of technological offerings from industry. Thus, there are fundamentally three issues that must be addressed by the strategic planners, namely:

- Funding issues – Requiring trade-offs to be made between the operational needs of the military and the budget constraints of the government.
- Requirements issues – Requiring trade-offs to be made between the budget constraints of the government and the technology availability from industry.
- Maturity issues – Requiring trade-offs to be made between the operational needs of the military and the technology availability from industry.

From a market-pull perspective, a key driver at the moment is the changing role of military forces. The type of future military operations has become significantly less predictable with the knock-on effect being the increasing need for defence forces to be able to change the focus of their capabilities as threats evolve (Churchill 2004). From a technology-push perspective, another key driver for change is the rate of technology development. Such change is characterised by both rapid advances in science and technology and shorter lead-times from research and development to certification (Kerr *et al.* 2008a). It is acknowledged that "effective defence capability is critically dependent on having up-to-date equipment and systems that incorporate the latest appropriate technology" (Bennett *et al.* 2004). The decision to introduce new product platforms (e.g. ships and submarines) is very high on the agenda of strategic planners. However, given the current state of defence budgets, there is a trend in sustaining the operational capability of current in-service platforms for much greater periods (Kerr *et al.* 2008a). Future plans must address this dynamic in order to manage the transition of the current portfolio (or fleet) of products to a mix of both new and existing systems. It necessitates the periodic upgrading of platform functionalities and associated performance levels of the in-service fleet. This directly translates to the introduction of new technology into legacy platforms to provide the needed capability for the swing in military operations (Kerr *et al.* 2010b). Thus there is the need to incorporate upgrades into platforms, as necessary, in order to respond to new threats (MoD 2005) and therefore the technology insertion activity has risen in importance within the defence community (Kerr *et al.* 2010b). It is within this context that the application of roadmapping must provide a clear and coherent visual summary of the Navy's strategic plans for realising the vision of '*Force 2030*'.

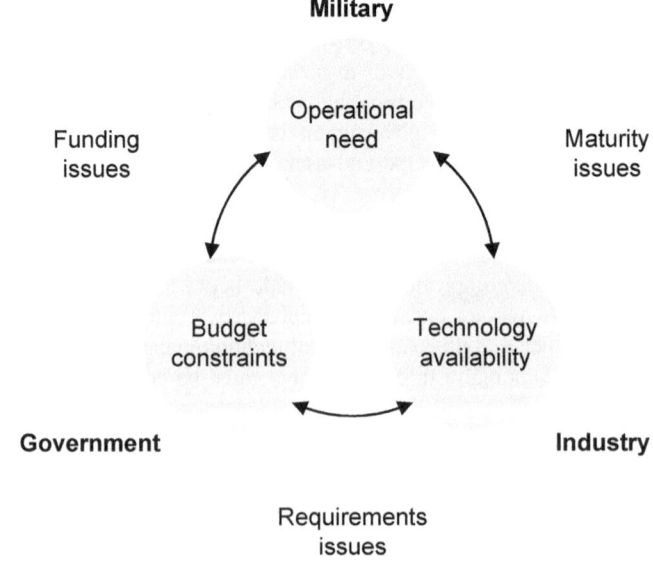

Fig. 1 Problem statement (Adapted from Kerr *et al.* 2010a)

3 Roadmapping Approach

Figure 2 gives a visual overview of the roadmap that was created to depict the Royal Australian Navy's contribution to realising the vision of *'Force 2030'*. This section will outline the approach taken in terms of the process for populating the roadmap together with the layout of the template. The underlying philosophy of the roadmap's development comes from the planning paradigm of Ackoff (1981) – specifically, the idea of *'ends planning'* whereby a desirable future is specified. Thus, strategic planning is "centred around the design of an idealised future" (Jackson 1990). In futures studies, this type of approach is termed *'normative'* (Saritas and Oner 2004). A normative approach asks the question: How can a specific target be reached? (Börjeson *et al.* 2006). In this case, the vision of *'Force 2030'*. It is worthy to note that this goal-orientated approach has much in common with the classic example of the Semiconductor Industry Association's (SIA) Technology Roadmap (1992a, 1992b), which is now commonly referred to as the International Technology Roadmap for Semiconductors (ITRS).

3.1 Process for Populating the Roadmap

To generate the roadmap presented in Figure 2, the normative approach necessitated an overall top-down process that started with the objectives stated in the Defence White Paper (DoD 2009a). Through the application of concept mapping, a backcast was conducted to plot the principal tasks, future roles and military capabilities required of the Navy in 2030. An end-state visual depiction of the maritime force elements was generated using the strategic capabilities-based representation developed by Kerr *et al.* (2008b). From this end-state, the platform transitions of the various vessels (ships and submarines) was then mapped taking into account the path dependencies of the legacy fleet. Finally, the main projects to be conducted in order to deliver newly upgraded systems to in-service vessels was plotted over the medium-term to account for capability shortfalls/gaps until the newly introduced platforms obtain their full operational readiness. This process for populating the roadmap is shown in Figure 3 and will now be described. Following the process description, the construction and design of the template will then be given in Section 3.2.

3.1.1 Top-Down

The planning associated with carrying out a policy falls within the area of policy implementation (Montjoy and O'Toole 1979); as opposed to the process that creates the policy vision in the first instance. The officials responsible for

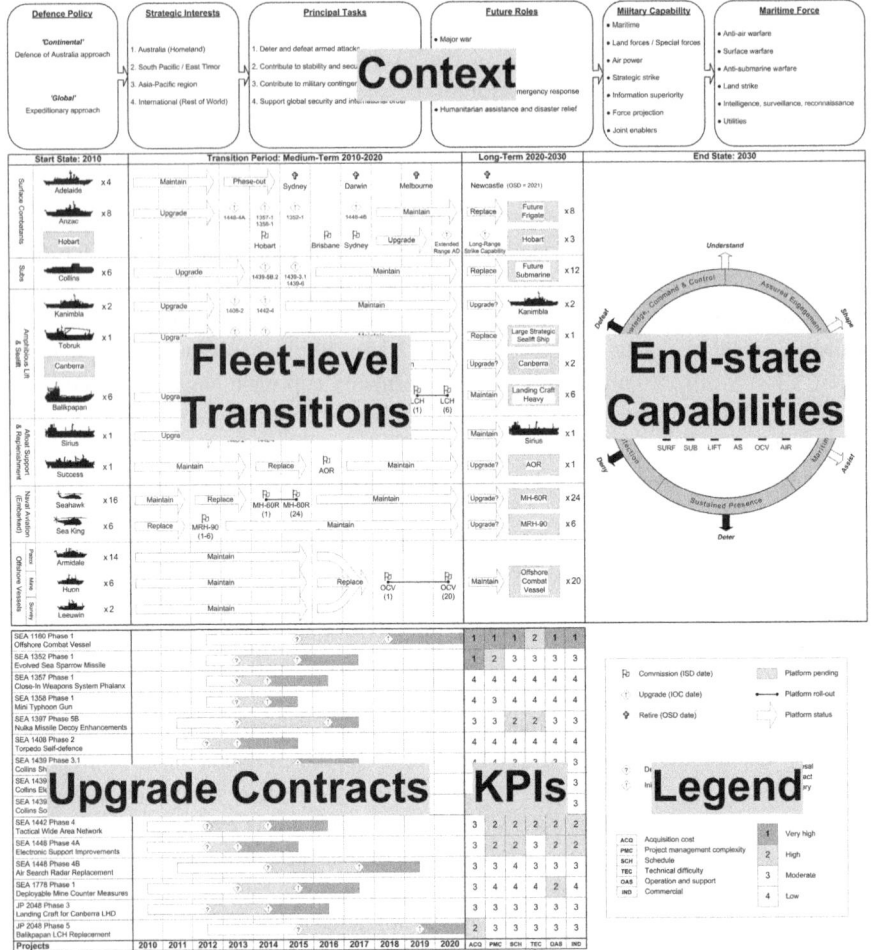

Fig. 2 Roadmap of the Navy's contribution to Force 2030

implementing policy are simply receptors of orders emanating from the vision (Crosby 1996). Authors such as Bardach (1977) have compared policy implementation to an assembly process which must be capable of producing the results called for. In this regard, planning and decision-making at this level is largely a top-down activity (Kahler 1989). Therefore in generating the Navy roadmap, the process is one of an overall top-down nature as depicted in Figure 3. From the field of public policy administration, the top-down orientated process is based on the body of work from Daniel Mazmanian and Paul Sabatier (Mazmanian and Sabatier 1983; Sabatier and Mazmanian 1979, 1980).

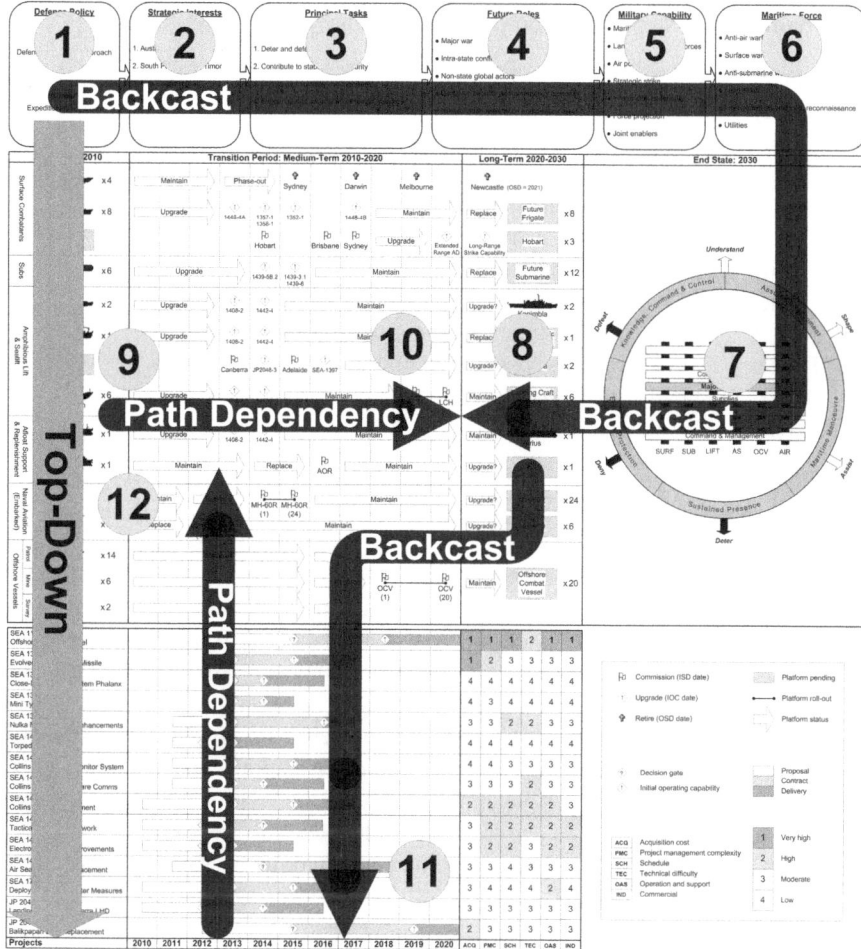

Fig. 3 Roadmapping process

They provide a set of necessary conditions for effective policy implementation; two of which have a strong influencing effect on promoting the relevance and applicability of roadmapping to the implementation planning challenge, namely:

- Clear and consistent objectives (van Meter and van Horn 1975) – These form the cornerstone of the roadmapping process as shown as Steps 1-6 in the top layer of Figure 3.
- Adequate causal theory (Pressman and Wildavsky 1973) – The canvas of a roadmap can readily accommodate a visual portrayal of causal linkages (Steps 1-12 in Figure 3).

The Navy's roadmap (Figure 2) was initiated by a mapping of the future strategic environment. The steps of the top-down analysis are as follows:

1. Depict the future context
    (i) Map the future conflicts
    (ii) Map the future roles
2. Depict the government's response
    (i) Map the strategic interests
    (ii) Map the policy stance
    (iii) Map the principal tasks
3. Depict the military's required capabilities
    (i) Map the key elements for the armed forces
4. Depict the Navy's required capability
    (i) Map the maritime vision
    (ii) Map the maritime force
    (iii) Map the maritime platforms

The rationale for utilising concept mapping as a precursor to populating the top layer of the roadmap was that "knowledge can sometimes be more easily elaborated once its skeletal form is revealed in visual form" (Huff and Jenkins 2002). It must be noted that there is a distinction being made in terminology; concept mapping was used since the source of data was from the government white paper whereas cognitive mapping signifies when an individual uses concept mapping to help clarify their own thinking, i.e. cognition (Bryson *et al.* 2004). Guidance on the technique can be found in the seminal text 'Mapping strategic thought' by Huff (1990) and the classic book written by Kane and Trochim (2007) entitled 'Concept mapping for planning and evaluation'. The output from concept mapping was further codified and translated into a causal flow diagram. Thus, the top-down analysis starting from the strategic environment provides the '*why*' context for the set of military capabilities that will keep Australia secure into the future.

3.1.2 Backcast

The top-down process through concept mapping was used to derive the '*why*' context by encapsulating the likely future strategic environment. To translate this material into a form suitable for populating the roadmap, a '*backcast*' was conducted and plotted onto the roadmap as illustrated in Figure 3. The idea of backcasting originated from Amory Lovins' (1976, 1977) '*soft energy path*' approach with its backwards-looking analysis. It involved describing a desirable future together with an assessment of how such a future could then be achieved (Anderson 2001). From a methodological perspective, the development of backcasting as a technique can be attributed to Robinson (1982). It was Robinson (1982) who translated Lovins' principles of '*looking backwards*' and '*exploring pathways*' into a formal method (Mander *et al.* 2008). Backcasting is an approach where "a vision or image of the future is first created, and then the series of steps needed to arrive there is worked out backwards, as it were, from the future end-state to the present-day starting point" (Voros 2006). There are three key attributes of backcasting that are of critical importance to the adoption and integration of the method onto the canvas of a roadmap, they are:

1. Normative
2. Design-orientated
3. Sequenced

Firstly, backcasting is explicitly normative (Höjer and Mattsson 2000; McDowall and Eames 2006; Robinson 1982, 1988, 1990, 2003; van Notten *et al.* 2003) because "the future is being chosen and not predicted" (Robinson 1990). In terms of asking the question of what future is desired (Höjer and Mattsson 2000), a backcast begins with a definition of the future goals and objectives (Robinson 1988). These are plotted in the top layer of the roadmap (labelled '*Context*' in Figure 2) across six points (Figure 3), namely:

- Point 1 – Plot the defence policy stance
- Point 2 – Plot the strategic interests
- Point 3 – Plot the principal tasks
- Point 4 – Plot the future roles
- Point 5 – Plot the military capabilities
- Point 6 – Plot the maritime force

Secondly, backcasting is design-oriented (Robinson 1988) because it is driven by an explicit image of the future (Robinson 1982) and there needs to be "a clear picture of the nature of the future being aimed at" (Robinson 1990). Mander *et al.* (2008) term this image as the desired end-point. In the case of the Navy's roadmap, such an end-point is given at Point 7 in Figure 3 and shows the end state maritime capabilities that are to be in place by 2030. It represents the vision for the Navy's contribution to '*Force 2030*' and defines the future warfighting elements, effects and complex product-service systems.

Thirdly, backcasting involves sequencing or reasoning from the specified future end state (van Notten *et al.* 2003). The question evolves from what future is desired to that of how can this desirable future be attained (Höjer and Mattsson 2000; Robinson 1982, 1990). Essentially it's about how to get to where you want to be? (Mander *et al.* 2008). It involves taking the desired end-point as the starting point and working back in time to the present in order to elicit what must be done to realise the vision (Mander *et al.* 2008; Robinson 1982, 1990). The analysis can be thought of as exploring and identifying the pathways to arriving at the desirable future situation (Dreborg 1996; McDowall and Eames 2006; Robinson 2003; van Notten *et al.* 2003). It is "a kind of reverse evolutionary form of thinking which starts with a vision and works backwards at the level of things-which-need-to-be-done or events-which-need-to-be-made-to-happen" (Voros 2006). The sequencing part of backcasting starts at Point 7 of Figure 3 and works back through two pathways:

- Point 7 to Point 8 – This considers the fleet-level transitions that must take place to arrive at the Navy's end state capability vision.
- Point 8 to Point 11 – This considers the technological-based projects that must be delivered into the fleet during the interim.

These pathways are focused on the new systems, i.e. the introduction of new naval vessels into service and breakthrough technologies. However, account must also be taken of the pathways related to the legacy systems. This perspective is treated in the next section where path dependency theory is applied to roadmapping.

3.1.3 Path Dependent

The future state of an organisation such as the Royal Australian Navy is heavily influenced by its current situation. Accordingly, it can be considered what Bruggeman (2002) terms as a path dependent organisation. The concept of path dependency originated from the economics field (Lynch 2006) to recognise that certain elements do constrain the future (Teece *et al.* 1997); as in the case of the Navy's investment in its current fleet of ships, submarines and helicopters. The significance of path dependency in resource commitment was established by both David (1985) and Arthur (1989). This concept "has played an important role in building the theoretical pillars of the resource-based view – the intellectual root of the capabilities approach" (Masrani and McKiernan 2009). The resource-based view of strategy development (Wernerfelt 1984) stresses the importance of the individual resources of the organisation in delivering competitive advantage" (Lynch 2006). Teece *et al.* (1997) acknowledged that "the notion of path dependencies recognises that history matters" and therefore "where a firm can go is a function of its current position and the paths ahead". Thus given the nature of the Navy's current fleet as a large fixed asset base, it will constrain future choices over the medium-term until the transition is made to newer vessels through the Department of Defence's procurement process. As noted by Davies (2008), "the Navy will look much as it does today until the middle of the next decade, when the air warfare destroyers and amphibious ships are delivered". To account for such a limiting factor, the roadmapping process has two pathways that deal with how the legacy fleet constrains the Navy's future over the period 2010-2020. They are shown in Figure 3 as:

- Point 9 to Point 10 – This considers the present day fleet as the start state and determines whether a particular class of vessel will be maintained, upgraded, replaced or phased-out.

- Point 11 to Point 12 – This considers the upgrade projects that will be necessary to sustain the operational capabilities and associated readiness levels of the systems on the older vessels.

3.2 Template for Presenting the Roadmap

A roadmap provides a visual canvas upon which a depiction of strategic plans can be articulated and shared both within and between organisations (Kerr *et al.* 2010c); but for it to effectively function as a boundary object, a roadmap should be developed taking into consideration graphical design principles for clear communication. The sub-domain of information design is specifically relevant to roadmapping. Information design focuses on "visually structuring and organising information to develop effective communication" (Watzman and Re 2008). "One of the great assets of graphical techniques is that they can convey large amounts of information in a small space" (Wainer 1984), therefore the graphical design intent for a roadmap should be to "condense a large amount of information into an intuitive format" (Ma *et al.* 2006). In roadmapping, the most fundamental graphical depiction is the single-page, high-level strategic view. According to Phaal and Muller (2009), such a condensed visual format of a roadmap constitutes a '*strategic lens*' on the problem. Phaal *et al.* (2008) recommend single-page visualisations as they ensure that "the key issues are focused on". Additionally, "one-page views can also be updated more easily, enabling the process to be more agile, enabling the roadmaps to keep pace with the rapidly changing business situations" (Phaal *et al.* 2008). Such '*single-pagers*' have a high degree of graphical equivalence with information dashboards (Kerr *et al.* 2010c); specifically, they share two key aspects:

- Dashboards must be customised – The design "must be tailored specifically to the requirements of a given person, group, or function; otherwise it won't serve its purpose" (Few 2006).
- The single-page presents a graphical challenge – "The first and toughest goal of a dashboard designer is to squeeze the information onto a single screen. All relevant information should be instantaneously viewable" (Eckerson 2006). It involves "squeezing a great deal of useful information and often disparate information into a small amount of space, all the while preserving clarity" (Few 2006).

In this regards, roadmaps can be a powerful graphical canvas for communicating strategic implementation plans by achieving an appropriate balance between the combination of two factors – (i) the structure that is embedded on the canvas and (ii) the appropriate set of visual objects that are then overlaid to represent informational content. Phaal and Muller (2009) were the first to explicitly recognise the distinction between the underlying information-based structure and the overlaying graphical style. This section will deal with the configuration of the layout, i.e. how the information contained within the roadmap is organised (Phaal and Muller 2009).

In management science, modelling involves framing the problem (Pidd 2003). Goffman (1974) introduced the term *'framing'* to refer to a scheme of interpretation, i.e. a framework, as a means to making sense of situations. From this perspective, roadmapping provides a "framework within which all the time-based business strategies of an enterprise can be aligned on a continuous basis in support of the business goals" (Whalen 2007). The framing in a roadmap refers to how information is organised on the canvas (Petrick and Provance 2005). Phaal *et al.* (2004b) terms this aspect of structuring a roadmap as its *'architecture'*. At its most basic level, the architecture of a roadmap consists of two dimensions: (i) the horizontal axis, which is most commonly a timeline, and (ii) the vertical axis, which consists of a number of layers that relates to how the business is conceptually or physically viewed (Phaal *et al.* 2004b) with each layer providing input into the next level (Cosner *et al.* 2007). According to Phaal and Muller (2009), a roadmap at its most elemental should be comprised of three broad layers. Probably the most profound framing mechanism for any roadmap is Kipling's wise men of *'Who, What, Where, When, Why and How'*. These come from a famous verse in Rudyard Kipling's 'The Elephant's Child' (part of the 'Just So Stories' collection):

> *I keep six honest serving-men*
> *(They taught me all I knew);*
> *Their names are What and Why and When*
> *And How and Where and Who.*

Phaal *et al.* (2004b) give a generic roadmap architecture consisting of three layers across the two-dimensional space of the canvas:

- Horizontal axis – Time, the *'know-when'* dimension
- Top layer – Purpose, the *'know-why'* dimension
- Middle layer – Delivery, the *'know-what'* dimension
- Bottom layer – Resources, the *'know-how'* dimension

This generic form was further elaborated upon by Phaal and Muller (2009), who state that the:

- Top layer – "Relates to the trends and drivers that govern the overall goals or purpose associated with the roadmapping activity".
- Middle layer – "Relates to the tangible systems that need to be developed to respond to the trends and drivers (top) layer. Frequently this relates directly to the evolution of products (functions, features and performance)".

- Bottom layer – "Relates to the resources that need to be marshalled to develop the required products, services and systems, including knowledge-based resources, such as technology, skills and competences and other resources such as finance, partnerships and facilities".

The generic architecture forms the starting basis for constructing the roadmap layout; in this case, Figure 4 shows the customised architecture that was generated for the Navy roadmap.

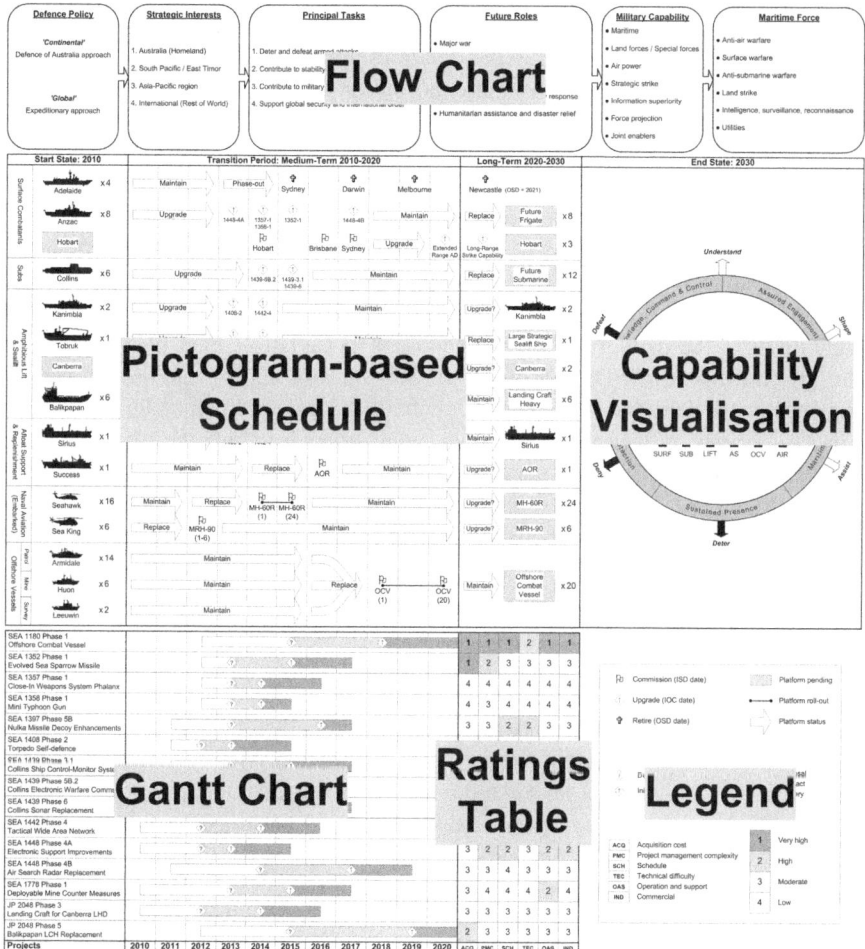

Fig. 4 Roadmap architecture

The top layer, which focuses on the '*know-why*' dimension, consists of a flow chart scheme. The flow chart is actually causal in the sense that there is a causal connection between the concepts (Jenkins 2002), i.e. "the arrows indicate how one idea or action leads to another" (Bryson *et al.* 2004). In Huff's (1990) categorisation of mapping, this type falls within the category of "influence, causality and system dynamics". This visual scheme is useful "where effective and logical communication is needed" (Bryson *et al.* 2004). Additionally, Phaal *et al.* (2004a) have identified that the pictorial representation of a flow chart is typically used in roadmaps to relate objectives, actions and outcomes. A good example is NASA's Strategic Management System Roadmap (NASA 1998) which used a flow chart scheme to illustrate the relationship between its vision, missions, fundamental scientific questions and contribution to national priorities. Thus, the top layer provides what Schön (1982) refers to as the contextual framing for reflection.

The middle layer, which focuses on the '*know-what*' dimension, is composed of two parts. On the right-hand side is the capability visualisation which is used to depict the end state vision for the Navy; whereas, on the left is the pictogram-based schedule which illustrates the fleet transitions from the start state (2010) over the medium- (2010-2020) and long-term (2020-2030) to arrive at the end state (2030). The pictogram-based schedule is essentially a fleet plan that shows the in-service dates (ISD) and out-of-service dates (OSD) of the various vessels together with delivery of important technology enhancements during the in-service periods. Given the top-down goal orientation of the roadmap, the capability visualisation is a key component of the template (Figure 4). It provides a visual summary of the Navy's strategic capabilities and acts as the linking point in the backcast down from the top layer (Figure 3). For transformation to be an effective process, there must be a clear and common understanding of the military capability required (Kerr *et al.* 2006). In order to provide a clear vision, the capability visualisation adopts the representation developed by Kerr *et al.* (2008b).

The bottom layer, which focuses on the '*know-how*' dimension, has two interconnected visual elements – Gantt chart and ratings table. This layer is the next level down in granularity and it aims to provide planners and programme managers with high-level metadata relating to the upgrade projects over the medium-term as the legacy fleet is phased-out and transitioned to newer classes. As such it is graphically aligned with the pictogram-based schedule directly above; such an architecture, of course, fully acknowledges that alignment is one of the universal principles of design (Lidwell *et al.* 2003). The Gantt chart in the bottom layer identifies the timing of the phases (proposal, contract, delivery) for each project. Alongside the Gantt chart is a ratings table which provides a scoring mechanism for each of the projects against six attributes (ANAO 2008; DoD 2009c), namely:

- Acquisition cost – What confidence do you have in the project cost estimate?
- Project management complexity – How well do you understand the solution?
- Schedule – How realistic is the schedule?
- Technical difficulty – What is the technical complexity in delivering the solution?
- Operation and support – What is the impact on the existing operating and support environment?
- Commercial – What confidence do you have that industry can deliver the solution?

4 Summary

A roadmap can be a very powerful visual medium for depicting strategic plans. When it is used to function as a boundary object, an appropriately constructed and designed single-page roadmap has the ability to convey a large amount of information in an intuitive format for communication both within and between organisations. This chapter demonstrated how roadmapping can be applied to the challenge of developing and displaying implementation plans in response to the goals articulated by government policy and how such future visions can be realised. The example given looked at the case of the Royal Australian Navy and how it could visualise its plans for contributing to the national defence force required to be in place by 2030 as embodied in the Australian Government's latest Defence White Paper. The content was focused on transitioning the fleet to fulfil this new vision and therefore the roadmap was orientated to the product platform view and associated technological-based project perspective.

In terms of process, the overall approach was normative, i.e. goal-orientated to realising the vision of '*Force 2030*'. Thus a top-down process was developed that started by concept mapping the future context, the government's response and strategic interests, the key elements for the armed forces, and then articulating the required contribution to military capability from the Navy. This information was plotted onto the canvas of the roadmap through backcasting in order to provide a causal linkage down from the defence policy stance and strategic interests, through to principal tasks and future roles, and then down to the military capabilities and maritime force elements. Additionally, given the Navy's current set of in-service assets, path dependencies were plotted to take into consideration the present day fleet to determine whether a particular class of vessel would be maintained, upgraded, replaced or phased-out. Also, consideration was made in regards to the upgrade projects that would be necessary to sustain the operational capabilities and associated readiness levels of the systems on the older vessels until the next generation replacements materialised.

In terms of graphical depiction, the roadmap canvas was constructed using a three layer architecture. The top layer representing the '*know-why*' dimension used a causal flow chart to display the strategic requirements. The middle layer representing the '*know-what*' dimension used a bespoke capability visualisation to display the end-state vision of the future naval force and a pictogram-based schedule to display the fleet transitions from the current start state over both the medium- and long-term. The bottom layer of the roadmap, representing the '*know-how*' dimension, used the combination of a Gantt chart and ratings table to display the projects constituting the technological resource-base. For each project, the Gantt chart gave the proposal, contract and delivery phases along with their associated decision gates; whereas, the ratings table provided measures for acquisition cost, complexity of project management, schedule constraints, technical difficulty, operation and support readiness, and commercial arrangements against each specific project.

It is hoped that this chapter provides practitioners with a guide through the method and its visual output, and additionally acts as a key reference case that demonstrates what can be achieved from adopting roadmapping.

References

Ackoff, R.L.: Creating the corporate future: Plan or be planned for. John Wiley and Sons, New York (1981)

ANAO – Australian National Audit Office, Defence Materiel Organisation: 2007-08 major projects report. Report Number: 9, Australian National Audit Office. Canberra, Australia (2008)

Anderson, K.L.: Reconciling the electricity industry with sustainable development: Backcasting, a strategic alternative. Futures 33(7), 607–623 (2001)

Arthur, B.: Competing technologies, increasing returns, and lock-in by historical events. Economic Journal 99(394), 116–131 (1989)

Bardach, E.: The implementation game. MIT Press, Cambridge (1977)

Bennett, L., Kinloch, A., McDermid, J., Muttram, R., Price, P., Stein, P., Stewart, W., Churchill, A., Jordan, G., Graeme-Morrison, B., Oxenham, D., Raby, N., Smith, H.: Key issues for effective technology insertion. Journal of Defence Science 9(3), 103–107 (2004)

Börjeson, L., Höjer, M., Dreborg, K.H., Ekvall, T., Finnveden, G.: Scenario types and techniques: Towards a user's guide. Futures 38(7), 723–739 (2006)

Bruggeman, D.: NASA: A path dependent organization. Technology in Society 24(4), 415–431 (2002)

Bryson, J.M., Ackermann, F., Edin, C., Finn, C.B.: Visible thinking: Unlocking causal mapping for practical business results. John Wiley and Sons, Chichester (2004)

Chandrasekaran, B., Kurup, U., Banerjee, B., Josephson, J.R., Winkler, R.: An architecture for problem solving with diagrams. In: The 3rd International Conference on the Theory and Application of Diagrams, Cambridge, United Kingdom, March 22-24 (2004)

Churchill, A.: The DSAC view on technology insertion. Distillation – The Science Journal for Dstl Staff 6, 5–8 (2004)

Clarke, A.E.: Situational analysis: Grounded theory after the postmodern turn. Sage Publications, Thousand Oaks (2005)

Cosner, R.R., Hynds, E.J., Fusfeld, A.R.: Integrating roadmapping into technical planning. Research-Technology Management 50(6), 31–48 (2007)

Crosby, B.L.: Policy implementation: The organizational challenge. World Development 24(9), 1403–1415 (1996)

David, P.: Clio and the economics of QWERTY. American Economic Review 75(2), 332–337 (1985)

Davies, A.: ADF capability review: Royal Australian Navy. Australian Strategic Policy Institute, Barton (2008)

DoD – Department of Defence, Defending Australia in the Asia Pacific century: Force 2030, Department of Defence, Canberra, Australia (2009a)

DoD – Department of Defence, Defence annual report: Volume 1. Department of Defence, Canberra, Australia (2009b)

DoD – Department of Defence, Defence capability plan. Department of Defence, Canberra, Australia (2009c)

Dreborg, K.: Essence of backcasting. Futures 28(9), 813–828 (1996)

Eckerson, W.W.: Performance dashboards: Measuring, monitoring and managing your business. John Wiley and Sons, Hoboken (2006)

Eppler, M.J., Platts, K.W.: Visual strategizing: The systematic use of visualization in the strategic-planning process. Long Range Planning 42(1), 42–74 (2009)

Few, S.: Information dashboard design: The effective visual communication of data. O'Reilly Media, Sebastopol (2006)

Gindy, N.N.Z., Cerit, B., Hodgson, A.: Technology roadmapping for the next generation manufacturing enterprise. Journal of Manufacturing Technology Management 17(4), 404–416 (2006)

Goffman, E.: Frame analysis: An essay on the organization of experience. Harper and Row, New York (1974)

Höjer, M., Mattsson, L.G.: Determinism and backcasting in future studies. Futures 32(7), 613–634 (2000)

Huff, A.S.: Mapping strategic thought. John Wiley and Sons, Chichester (1990)

Huff, A.S., Jenkins, M.: Mapping strategic knowledge. Sage Publications, London (2002)

Jackson, M.C.: Russell Ackoff's Jerusalem. Systems Practice 3(2), 177–182 (1990)

Jenkins, M.: Cognitive mapping. In: Partington, D. (ed.) Essential Skills for Management Research, pp. 181–198. Sage Publications, London (2002)

Kahler, M.: International financial institutions and the politics of adjustment. In: Nelson, J.M. (ed.) Fragile Coalitions: The Politics of Economic Adjustment, pp. 139–160. Overseas Development Council, Washington (1989)

Kane, M., Trochim, W.M.K.: Concept mapping for planning and evaluation. Sage Publications, Thousand Oaks (2007)

Kerr, C., Phaal, R., Probert, D.: A framework for strategic military capabilities in defense transformation. In: The 11th International Command and Control Research and Technology Symposium (ICCRTS 2006) - Coalition Command and Control in the Networked Era, Cambridge, United Kingdom, September 26-28 (2006)

Kerr, C.I.V., Phaal, R., Probert, D.R.: Technology insertion in the defence industry: A primer. Proceedings of the Institution of Mechanical Engineers, Part B: Journal of Engineering Manufacture 222(8), 1009–1023 (2008a)

Kerr, C., Phaal, R., Probert, D.: A strategic capabilities-based representation of the future British armed forces. International Journal of Intelligent Defence Support Systems 1(1), 27–42 (2008b)

Kerr, C.I.V., Phaal, R., Probert, D.R.: Cogitate, articulate, communicate: The psychosocial reality of technology roadmapping and roadmaps. In: The R&D Management Conference 2009 – The Reality of R&D and its Impact on Innovation, Vienna, Austria, June 21-24 (2009)

Kerr, C., Phaal, R., Probert, D.: Inserting Innovations In-Service. In: Finn, A., Jain, L.C. (eds.) Innovations in Defence Support Systems – 1. SCI, vol. 304, pp. 17–53. Springer, Heidelberg (2010a)

Kerr, C.I.V., Phaal, R., Probert, D.R.: Ranking maritime platform upgrade options. Proceedings of the Institution of Mechanical Engineers, Part M: Journal of Engineering for the Maritime Environment 224(1), 47–59 (2010b)

Kerr, C., Phaal, R., Probert, D.: Depicting options and investment appraisal information in roadmaps. In: The 19th International Conference on Management of Technology (IAMOT 2010) – Technology as the Foundation for Economic Growth, Cairo, Egypt, March 8-11 (2010c)

Lidwell, W., Holden, K., Butler, J.: Universal principles of design. Rockport Publishers, Gloucester (2003)

Lovins, A.B.: Energy strategy: The road not taken? Foreign Affairs 55(1), 65–96 (1976)

Lovins, A.B.: Soft energy paths: Toward a durable peace. Penguin Books, London (1977)

Lynch, R.: Corporate strategy, 4th edn. Pearson Education, Harlow (2006)

Ma, T., Liu, S., Nakamori, Y.: Roadmapping as a way of knowledge management for supporting scientific research in academia. Systems Research and Behavioral Science 23(6), 743–755 (2006)

Mann, W.C.: Dialogue games: Conventions of human interaction. Argumentation 2(4), 511–532 (1988)

Mander, S.L., Bows, A., Anderson, K.L., Shackley, S., Agnolucci, P., Ekins, P.: The Tyndall decarbonisation scenarios: Part 1 development of a backcasting methodology with stakeholder participation. Energy Policy 36(10), 3754–3763 (2008)

Masrani, S.K., McKiernan, P.: Addressing path dependency in the capabilities approach: Historicism and foresight meet on the 'road less travelled'. In: Costanzo, L.A., MacKay, R.B. (eds.) Handbook of Research on Strategy and Foresight, pp. 485–504. Edward Elgar, Cheltenham (2009)

Mazmanian, D.A., Sabatier, P.A.: Implementation and public policy. Scott Foresman and Company, Glenview (1983)

McDowall, W., Eames, M.: Forecasts, scenarios, visions, backcasts and roadmaps to the hydrogen economy: A review of the hydrogen futures literature. Energy Policy 34(11), 1236–1250 (2006)

MoD – Ministry of Defence, Defence industrial strategy. Report Number: Cm 6697, The Stationery Office, London, United Kingdom (2005)

Montjoy, R.S., O'Toole, L.J.: Toward a theory of policy implementation: An organizational perspective. Public Administration Review 39(5), 465–476 (1979)

NASA – National Aeronautics and Space Administration, NASA strategic plan. Report Number: NPD-1000.1a, National Aeronautics and Space Administration, Washington DC, United States of America (1998), http://www.hq.nasa.gov/office/codez/plans.html

Petrick, I.R., Provance, M.: Roadmapping as a mitigator of uncertainty in strategic technology choice. International Journal of Technology Intelligence and Planning 1(2), 171–184 (2005)

Phaal, R., Muller, G.: An architectural framework for roadmapping: Towards visual strategy. Technological Forecasting and Social Change 76(1), 39–49 (2009)

Phaal, R., Farrukh, C.J.P., Probert, D.R.: Technology roadmapping – A planning framework for evolution and revolution. Technological Forecasting and Social Change 71(1-2), 5–26 (2004a)

Phaal, R., Farrukh, C., Probert, D.: Customizing roadmapping. Research-Technology Management 47(2), 26–37 (2004b)

Phaal, R., Simonse, L., den Ouden, E.: Next generation roadmapping for innovation planning. International Journal of Technology Intelligence and Planning 4(2), 135–152 (2008)

Pidd, M.: Tools for thinking: Modelling in management science, 2nd edn. John Wiley and Sons, Chichester (2003)

Pressman, J., Wildavsky, A.: Implementation. University of California Press, Berkeley (1973)

Robinson, J.B.: Energy backcasting: A proposed method of policy analysis. Energy Policy 10(4), 337–344 (1982)

Robinson, J.B.: Unlearning and backcasting: Rethinking some of the questions we ask about the future. Technological Forecasting and Social Change 33(4), 325–338 (1988)

Robinson, J.: Futures under glass: A recipe for people who hate to predict. Futures 22(8), 820–842 (1990)

Robinson, J.B.: Future subjunctive: Backcasting as social learning. Futures 35(8), 839–856 (2003)

Sabatier, P., Mazmanian, D.: The conditions of effective implementation: A guide to accomplishing policy objectives. Policy Analysis 5(4), 481–504 (1979)

Sabatier, P., Mazmanian, D.: The implementation of public policy: A framework of analysis. Policy Studies Journal 8(4), 538–560 (1980)

Saritas, O., Oner, M.A.: Systemic analysis of UK foresight results: Joint application of integrated management model and roadmapping. Technological Forecasting and Social Change 71(1-2), 27–65 (2004)

Schön, D.A.: The reflective practitioner: How professionals think in action. Basic Books, New York (1982)

SIA – Semiconductor Industry Association, Semiconductor technology workshop conclusions. Semiconductor Industry Association, San Jose, United States of America (1992a), http://public.itrs.net

SIA – Semiconductor Industry Association, Semiconductor technology workshop working group reports. Semiconductor Industry Association, San Jose, United States of America (1992b), http://public.itrs.net

Star, S.L., Griesemer, J.: Institutional ecology, translations and boundary objects: Amateurs and professionals in Berkeley's museum of vertebrate zoology 1907-39. Social Studies of Science 19(3), 387–420 (1989)

Teece, D.J., Pisano, G., Shuen, A.: Dynamic capabilities and strategic management. Strategic Management Journal 18(7), 509–533 (1997)

van Notten, P.W.F., Rotmans, J., van Asselt, M.B.A., Rothman, D.S.: An update scenario typology. Futures 35(5), 423–443 (2003)

van Meter, D.S., von Horn, C.E.: The policy implementation process: A conceptual framework. Administration and Society 6(4), 445–488 (1975)

Voros, J.: Introducing a classification framework for prospective methods. Foresight 8(2), 43–56 (2006)

Wainer, H.: How to display data badly. The American Statistician 38(2), 137–147 (1984)

Walton, D.: Informal logic: A pragmatic approach, 2nd edn. Cambridge University Press, New York (2008)

Watzman, S., Re, M.: Visual design principles for usable interfaces: Everything is designed, why we should think before doing. In: Sears, A., Jacko, J.A. (eds.) The Human-Computer Interaction Handbook – Fundamentals, Evolving Technologies and Emerging Applications, 2nd edn., pp. 329–353. Lawrence Erlbaum Associates, New York (2008)

Wernerfelt, B.: A resource-based view of the firm. Strategic Management Journal 5(2), 171–180 (1984)

Whalen, P.J.: Strategic and technology planning on a roadmapping foundation. Research-Technology Management 50(3), 40–51 (2007)

Authors

Clive Kerr joined the Centre for Technology Management at the University of Cambridge as a Research Associate in April 2005. His current research interests are visual strategy, roadmapping, technology intelligence, technology insertion and through-life capability management. Prior to joining Cambridge, he was a Research Officer in Engineering Design at Cranfield University. Clive has a First Class Honours degree in Electrical and Mechanical Engineering co-awarded together with a Diploma in Industrial Studies, a Diploma degree in Economics, a Postgraduate Certificate in the Social Sciences and a Doctorate in Engineering. He is a Chartered Engineer with professional memberships of the IMechE, IET, RAeS and the AIAA.

Robert Phaal is a Principal Research Associate in the Engineering Department of the University of Cambridge, based in the Centre for Technology Management, Institute for Manufacturing. He conducts research in the area of strategic technology management, with a particular interests in technology roadmapping and evaluation, emergence of technology-based industry and the development of practical management tools. Rob has a mechanical engineering background, with a PhD in computational mechanics, with industrial experience in technical consulting, contract research and software development.

David Probert is a Reader in Technology Management and the Director of the Centre for Technology Management at the Engineering Department of the University of Cambridge. His current research interests are technology and innovation strategy, technology management processes, industrial sustainability and make or buy, technology acquisition and software sourcing. David pursued an industrial career with in the food, clothing and electronics sectors for 18 years before returning to Cambridge in 1991.

Part 2: Processes of Technology Roadmapping

Many processes have been proposed for developing technology roadmap(s), in a range of application contexts. The basic dichotomy of exploratory and normative-driven approaches provides an initial way of positioning roadmapping initiatives. Exploratory approaches can be differentiated further to define a triangle of market driven, technology- driven and normative-driven approaches. In addition, these processes may differ according to the level of application; technology roadmapping processes in networks will normally differ from those in firms. Six key process models are introduced in this part of the book, illustrating a variety of processes across these dimensions:

- *Phaal, Farrukh and Probert* focus on a process model for 'fast-start' technology roadmapping, based on multifunctional workshops. Two approaches are described, suitable for application at product and business level to support innovation and strategy. Product level technology roadmaps depend on a keen understanding of customer needs and market conditions. Business level initiatives often involve large groups of stakeholders, considering both the broad organizational context and specific opportunities and issues of interest.
- *Schuh, Wemhoener and Orilski's* technology-driven view takes as its starting point the evolutionary trajectory of a technical system, the most famous of which is Moore's law for the development of semiconductors. In this case the roadmapping method is applied as a management tool which enables enterprises to systematically identify and evaluate technologies, aligning technologies with business strategies. This roadmapping method shows how enterprises can successfully identify and use market opportunities at an early stage.
- *Geschka and Hahnenwald* adopt a market-driven view as the starting point for explorative technology roadmapping, based on environmental scenarios. In this process model the well-known scenario technique is applied to build an understanding of possible future situations. Based on this, technical ways of achieving these situations are developed, supporting strategic technology planning in the firm. The approach of Geschka and Hahnenwald gives decision makers a tool to react to changing external influencing factors that determine the development of a given technology.
- *Moehrle* provides a contrasting technology-driven perspective of TRIZ based technology roadmapping. This approach benefits from exploiting the field of technologically inspired opportunities, adding a market based view in a later stage.

- *Kanama* integrates different views, market- as well as technology-driven, in a Delphi based technology roadmapping approach. In this process model, results of a Delphi process (as often performed by a state agency such as NISTEP in Japan) form the starting point for technology roadmapping. Kanama discusses the strengths and weaknesses of the Delphi and technology roadmapping methods, and how a new approach can take advantage of the strengths of both methods.
- *Abe* presents a comprehensive business-oriented process model for normative technology roadmapping. The starting point is a vision of what can and should be achieved in the future, not necessarily a prognosis of the future. His Innovation Support Technology (IST) model integrates business modeling and roadmapping methods, supporting enterprises decision makers to increase corporate value through better exploitation of R&D outputs.

These six process models illustrate that there are different ways of creating roadmaps. For companies or decision makers this means that they need to know which visions of the future they want to develop, e.g. a technology-driven or more commercially driven vision (see Fig 1). The process has to be selected and/or adapted to suit the particular business and industry context.

The contributions are positioned against the framework shown in figure 1. Exploratory oriented process models encourage the identification and development of further opportunities, while goal directed approaches enable strategic planning at a more detailed level. Technology-oriented process models enable exploitation opportunities to be explored, while market-oriented approaches help to ensure that appropriate technological capability is available. Roadmapping is a knowledge intensive activity, encouraging dialogue between commercial and technical groups, as a learning process.

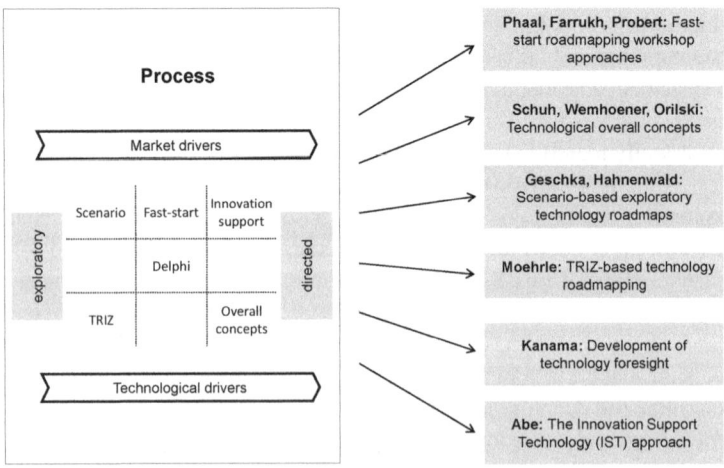

Fig. 1 Processes for successful technology roadmapping

Fast-Start Roadmapping Workshop Approaches

Robert Phaal, Clare Farrukh, and David R. Probert

Roadmaps support strategic alignment through the creation and use of structured visual representation of the various perspectives needed for successful business and innovation. The process of roadmapping is crucial to achieving this goal, as a means of involving the various stakeholders. Workshops often form a key element of the roadmapping process, providing an opportunity for participants to share and capture views in a creative environment. Workshop methods for supporting such dialogue are described, focusing on product (innovation) level roadmapping. A reference schedule is presented as a basis for designing workshops that bring together cross-functional groups to create a first draft roadmap in a short period of time, leading to consensus-based actions.

1 Introduction

This paper focuses on 'fast-start' workshop techniques, including the 'T-Plan' method for product-technology roadmapping, and the more general 'S-Plan' approach for strategy and policy applications. These approaches use interactive workshops to bring together diverse groups of participants to capture and discuss perspectives, focus and explore options and opportunities, make decisions and agree actions, and to develop preliminary roadmaps (a typical workshop is shown in Fig. 1). The methods have been developed over a period of 10 years, involving more than 200 wide-ranging collaborative applications in industry, government and academia.

There are two principal variants of the fast-start approach that apply similar concepts and techniques to address different application classes, although each case requires some degree of customisation depending on goals and context:

- *S-Plan* focuses on general strategic challenges, typically at business, corporate, sector and policy levels. The process brings together large groups of diverse stakeholders in 1-2 day workshops to explore and prioritise strategic issues, develop and align innovation and research strategies, and to agree the way forward (Phaal et al., 2007).

Robert Phaal · Clare Farrukh · David R. Probert
Centre for Technology Management, Institute for Manufacturing,
Department of Engineering, University of Cambridge
17 Charles Babbage Road, Cambridge, CB3 0FS,
United Kingdom
e-mail: {rp108,cjp22,drp1001}@cam.ac.uk

Fig. 1 Typical roadmapping workshop activity

- *T-Plan* focuses on product-technology roadmapping, bringing together medium sized groups of cross-functional stakeholders in four half-day workshops to explore and plan a product-based innovation (Phaal et al., 2001). The T-Plan process is described in detail in Section 3, including a case example.

The two methods may be combined, although either method can be used in isolation, including particular modules, or as part of other business processes. S-Plan has the capacity for covering a very broad scope (typically business, corporate or sector levels), rapidly capturing perspectives, identifying and prioritising key topics for exploration and action planning. If one of the topics relates to product, service or process level innovation, then T-Plan provides a more detailed method for developing aligned market, product, technology and resource strategies and plans. S-Plan can be used as a starting point, associated with the left-hand side of the funnel, with T-Plan more suited to a slightly later phase (second iteration), once there is confidence on where to focus innovation efforts.

2 Fast-Start Workshop Methods

The fast-start approach uses multi-functional and multi-organisational workshops as a means for the rapid initiation of roadmapping. The methods are designed to be agile, in the sense of being flexible, rapid, efficient and scaleable. Focusing on immediate issues of concern and interest delivers quick benefits, while taking the first step on what can be a long roadmapping journey. The main outputs of the first iteration in this rapid prototyping approach are decisions and agreed actions, together with process learning from a pilot application of roadmapping. The general fast-start workshop approach is summarised in this section, with more detail of the T-Plan approach provided in Section 3, including examples.

2.1 Role of Fast-Start Workshops

The way in which a roadmapping initiative supports strategy, policy or innovation in the organisation should be considered carefully during the design phase, as inputs and outputs to and from the roadmapping activity will often be linked to milestones within these business processes, as indicated in Fig. 2 (for example, review points in a new product development process). Individual workshops support the broader roadmapping initiative, as a microcosm of the overall process, requiring planning, implementation and follow-on, with inputs and outputs aligned with the overall roadmapping initiative. Support is required in terms of facilitation, steering and project management, and further work is typically needed before, between and after workshops to collect data, analyse results, develop roadmap representations and associated reports, and to ensure that actions are taken forward.

There is no single universally applicable roadmapping method, with a need to adapt and customise the approach to suit the particular circumstances. Roadmapping initiatives can be separated into three broad elements: preparation, implementation and follow-on. Also shown in Fig. 3 are the key success factors identified by de Laat & McKibben (2003) for supra-company (network / sector) level applications, most of which also apply to company initiatives.

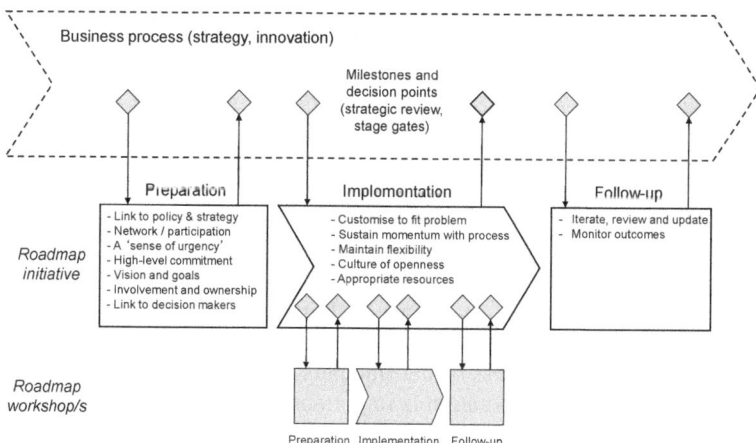

Fig. 2 Position of roadmapping initiatives and workshops within strategy and innovation processes, highlighting key success factors identified by de Laat & McKibben (2003)

Fast-start roadmapping workshop techniques enable key stakeholders to address strategic issues efficiently using the visual structure of roadmaps to capture, discuss, prioritise, explore and communicate issues. Focusing on current concerns, using the approach as a problem-solving tool, improves the likelihood of quick benefits, and provides a valuable learning opportunity and first step towards a sustainable process.

Roadmapping used in this rapid prototyping way can act as a useful diagnostic tool – mapping available knowledge and views will very quickly identify gaps in knowledge, together with issues and risks that require action. This 'agile' approach avoids the danger of over complicating and bureaucratising the process, which is a common pitfall and a reason why many roadmapping processes flounder. Moving forward, individual roadmaps, and the roadmapping system overall, can be developed as appropriate for the organisation, in terms of approach and the degree of formality required, to fit with its structure, culture and business processes.

When designing a roadmapping process it is important to consider the desired qualities of the output (the roadmap): what constitutes a 'good' roadmap? how should 'quality' be assessed? Clearly, confidence and accuracy of commercial and technological forecasts are helpful, and it is important to use the best available knowledge, information and expertise. But, recognising that such forecasts are often highly uncertain in the longer term and that there will be may gaps and questions associated with the first versions of the roadmaps produced, other outputs of the process provide a better measure of utility – primarily the decisions, consensus and actions that arise.

Treating a roadmap as a fixed project plan is dangerous, unless the purpose is to govern a set of projects to implement a complex programme. Rather, a roadmap should be considered as a type of 'radar', looking forward in order to improve understanding, enhance communication, build networks, capture knowledge, make decisions, agree priorities and take actions, steering the organisation into the future.

The S-Plan and T-Plan methods provide reference processes that can and should be adapted as required, based on a clear understanding of the issues being addressed. This can involve incorporation of other tools and frameworks, such as valuation techniques or scenario planning. Modules within the methods can be used independently to support other processes and workshops if helpful.

2.2 Planning a Fast-Start Workshop

It is important to start planning well in advance of the workshop. This includes a collaborative design process, as shown in Fig. 3, involving both the roadmap owner (the person or group who want to use roadmapping to address their strategic issues) and the process owner (the person or group who will manage and facilitate the process). Collaboration is important due to the need to customise the roadmapping approach to address the particular organisational context and goals.

Management and governance of the process should be considered. In companies this will usually involve a small team of senior managers, together with the process facilitators, to steer and review progress and outcomes, while for sector-level foresight initiatives more formal mechanisms may be required.

Fast-Start Roadmapping Workshop Approaches

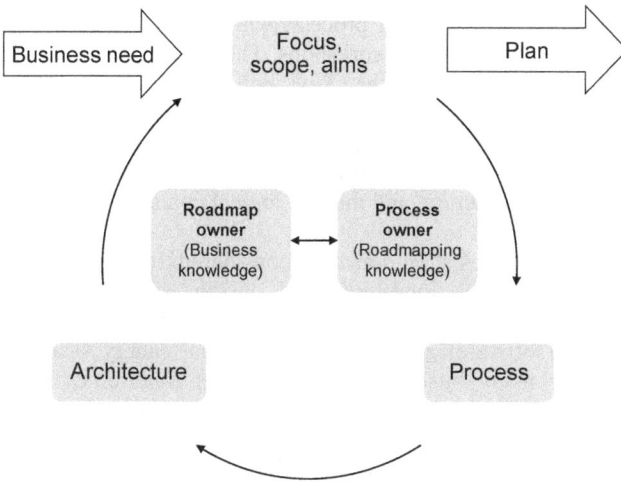

Fig. 3 Planning for roadmapping involves collaborative design

It is important to define objectives, focus and boundaries, to design the roadmap architecture and process, and to plan the logistics of the event. Key issues to consider include:

- What are the critical goals?
- What questions and issues are important to address?
- What are likely to be the most interesting and important topics?
- What timeframes need to be considered?
- What is already known?
- What other processes, methods and systems should the initiative link to?
- What might the outputs look like? Will these meet the aims?
- What unit of analysis is appropriate (a balance between breadth and depth)? What is the scope and where are the boundaries?
- How should the topic be structured?

Defining the architecture of the roadmap is a key step, as this provides the structure that guides virtually all activities in the process and workshop, providing a common visual language. The roadmap structure captures (and constrains) boundaries and timescales, allows pre-population with existing information and enables efficient population of roadmap in the workshop and subsequent reporting. The following six fundamental questions can be useful for designing the roadmap architecture, which relate to the horizontal (time) and vertical (layered) axes, respectively:

1a) Where do we want to go?   1b) Where are we now?   1c) How can we get there?

2a) Why do we need to act?   2b) What should we do?   2c) How should we do it?

For S-Plan the architecture of the roadmap is generally designed prior to the workshop, as part of the planning process, while for T-Plan the structure of the roadmap is developed during the workshops. The layers and sub-layers need to be defined, to represent all of the key perspectives and system elements appropriately, with the amount of space devoted to each layer reflecting the relative importance and likely density of information. It is sensible to 'test drive' the architecture before use in a workshop, confirming that key content can be positioned within the structure, and that the logical connections between the layers can be expressed.

Appropriate timeframes should also be considered, including the past, current, short-term, medium-term, long-term and vision. The actual time that is represented depends on the industry and associated rate of change. Roadmaps of between 2 years (software) and 100 years (national energy systems) have been observed, but for many businesses the following timeframes are appropriate: short-term 1-year budget horizon; medium-term 3-year strategy horizon; and long-term 10-year 'radar' horizon. A non-linear scale is often used to represent these time horizons, with more space devoted to the short and medium term, where the density of information is likely to be greatest.

The quality of the roadmap is largely dictated by the breadth and depth of expertise and knowledge of the participants in the workshop, and so careful consideration should be given to this during the planning phase. It is important to have a healthy mix of commercial and technical representation, and external views are helpful. The process benefits from diversity of perspectives, and the facilitation techniques need to be able to deal with the potential conflicts that may arise.

The structure of the roadmap provides an initial checklist for ensuring that all important perspectives are represented in the workshop. For company applications this will generally include participation from business, market, application and technology functions. Senior management support is vital if the process is to have impact, and to ensure participation is given appropriate priority. Workshop dates may be dictated by key participants' diaries, often resulting in lead-times of 4-8 weeks. Participants should be sent joining instructions, briefing material and any required pre-work approximately two weeks prior to the workshop.

For sector level foresight applications a range of industrial, academic and government participation is typical. Recruitment of workshop participants is more of a challenge for these situations, and considerable effort may be required to ensure success. Suitable participants should be identified early in the process, and as a rule of thumb one can expect 30-50% recruitment success, with personal contacts and follow up communication improving the response. It can be expected that a few participants will not turn up on the day, and so it is prudent to plan for several additional participants for this contingency.

A suitable venue is required, with the key requirement being sufficient space for activities, including breakout sessions. Plenty of floor and wall space is needed to accommodate the interactive nature of the workshops. For the plenary sessions the room should ideally be approximately double the size required for a traditional conference meeting involving the same number of participants. It is helpful if the room is reconfigurable, allowing tables and chairs to be moved for different activities. The room should be arranged with tables in either a 'U' shape or as 'islands', to facilitate interaction. It is advisable to view the room arrangements prior to the workshop, and to book the venue well in advance. It is essential that charts can be stuck up on the walls – if the venue does not allow this, suitable poster-boards should be sourced.

Roadmap-inspired structures are used to guide the overall process and most of the workshop activities, although other tools and frameworks can be used to enhance or supplement the method. For example, a portfolio selection matrix can support topic prioritisation if appropriate. Professional printers can be used to produce the workshop charts, in colour if useful – allow enough time for printing and checking templates, as these are critical for the workshop process. Alternatively, charts can be created using flip chart paper, sticky tape and pens, which ensures flexibility during workshops if there is a need to refocus or adapt the process.

In addition to structured templates, a range of other stationery is required for roadmapping workshops:

- A plentiful supply of sticky notes, in a range of sizes, shapes and colours. Colours can be used for specific purposes (for example, if different business units are represented); although the main reason for using a range of colours is to help participants to navigate the content on the charts more easily. Different shapes (for example, arrows) are helpful for highlighting specific topics and issues during workshop activities.
- Sticky dots are useful for quick voting processes. Again, colour can be helpful, for example to rank options in terms of both reward and risk.
- Sticky tape, to construct charts from flip chart paper if required, and to secure all sticky notes at the end of the workshop.
- Masking tape or other devices to stick charts on walls.
- Miscellaneous stationery, such as flip chart paper, scissors and felt-tip pens for ease of writing on sticky notes.

2.3 Running a Fast-Start Workshop

The facilitation approach is fairly light, in the sense that the main focus of the workshop is on group-based activities, where it is the participants' experience and knowledge that is key, with interaction supported by the provision of structured frameworks (charts), clear steps and the means to capture, share and organise perspectives (sticky notes). The overall workshop agenda needs to be designed to meet the agreed aims, with the time available broken down into logical steps.

The ways in which sessions are facilitated, and what can be achieved within a given time period, depend on the size of the group:

- Small groups (< 5) can generally self-organise, with fairly minimal facilitation support. There is a trade-off between coherence and diversity. One person can draw a very neat roadmap, but it is only their view. Incorporating additional perspectives is hugely valuable for addressing complex and open-ended topics, although it becomes much more of a challenge to develop a

coherent and agreed output. Small groups can be relied on to deliver a result, with periodic review and support if required. Groups of 5 or more will find it difficult to proceed without active facilitation or guidance.
- Medium sized groups (5 ~ 10) require facilitation, supported by some knowledge of the topic of interest to enable active engagement. For this size of group the roadmap framework enables the topic to be tackled in a structured and logical way, with discussion guided by the facilitator and captured on wall charts.
- Large groups (> 10) require more formal and specialised facilitation techniques, with the activities orchestrated and managed professionally, often by a team of facilitators for workshops with 15-20 participants or more. Large group activities can be somewhat mechanistic, and so should be balanced appropriately with smaller group activities.

From a facilitator's perspective, timing is a key aspect to manage. Such workshops are very interactive and intense, and owing to the complex issues being considered and the large number of participants it is very easy for sessions to overrun. Larger groups invariably take longer to perform activities. It is vital that tight control of timing is maintained, while at the same time allowing for some flexibility, due to the inherently exploratory nature of roadmapping. It is vital that a satisfactory conclusion is reached by the end of the workshop, and every effort should be made to finish on time.

Considering these issues during the planning phase allows for contingencies to be built into the process. Breaks in the agenda provide natural review points to ensure that the process is on track, and if required, short unscheduled breaks can be used to address key issues that arise. Judgement must be exercised concerning how much time can be devoted to discussion of issues as they arise. Once the nature of the issue is clear, if it is not critical to the logic of the workshop process, then it is generally advisable to note the issue on a flip chart sheet maintained for this purpose, for review at a later stage. Be aware that the pace of activities tends to be slower at the start, as participants familiarise themselves with the process, and then to speed up.

Having two facilitators available is very helpful, in terms of splitting and rotating roles, allowing more complex group activities to be managed. Having a small supporting team available is useful, to review progress during breaks and to deal with unanticipated issues that may arise. This team would typically include the organisational sponsors of the roadmapping process, and /or their representatives, together with the facilitation team.

2.4 After a Fast-Start Workshop

An immediate task after the workshop is to transcribe all of the outputs as a full record of the event, and to interpret and summarise the outputs to create a report for circulation to participants for comment. For companies, it is recommended that the group work be captured and summarised by the groups themselves, as an immediate action, using templates provided for consistency, while for sector level applications the facilitation team would normally undertake this task. It is recommended that presentation software (e.g. PowerPoint) be used for reporting in the first instance, as this encourages an appropriate level of synthesis and summary, content can be extracted for presentation purposes, and it can be used as a resource for more formal reporting if required. The report can also be circulated and presented to a wider group for validation and dissemination purposes.

It should also be recognised that the workshop is one (critical) step within a wider process, even if it is being used as a one-off problem solving technique. Consideration should be given to the actions required to move forward and whether and how the roadmap will be updated as an element of this process (part of the final review discussion during workshops). For companies this is often linked to other core business processes, such as an annual strategy / budget cycle, or review points within research or new product development processes.

3 Product Level Roadmapping (T-Plan)

The T-Plan approach focuses on integrated product-technology strategic planning. The process brings together 8-12 participants from across the organisation to develop a draft roadmap for a product (or product family), in four half-day workshops:

1) *Market:* external market and internal business drivers are identified, categorised and prioritised for key market segments. Business strategy is reviewed and knowledge gaps identified.
2) *Product:* potential product features, functions and attributes are identified and prioritised with respect to how strongly they address the drivers. Product strategy is reviewed and knowledge gaps identified.
3) *Technology:* potential technological solutions for developing the product features are identified and prioritised and knowledge gaps identified.
4) *Charting:* based on the outputs from the first three workshops, the initial roadmap is developed, linking market, product and technology perspectives, decisions are made and actions agreed.

The T-Plan process is summarised below, including two company examples to illustrate the approach. For detailed guidance, including facilitation aspects, the reader is referred to the T-Plan guide (Phaal et al., 2001).

3.1 T-Plan Process

The underlying roadmapping principles are common to both the S-Plan and T-Plan processes, although the focus is more specifically at the product level (including associated market and technology perspectives). The main differences are:

- The focus on a single product or closely related product family requires greater granularity than for S-Plan, particularly in terms of market drivers (customer purchase motivations) and product functionality, performance and features. For large-scale complex products an S-Plan type approach may be appropriate in the first instance.
- Workshop participants should be selected for their expertise in relation to the product area of interest. Typically this might involve 8-12 experts, including commercial and technical perspectives, with participants taking part in all four workshops.
- The process is separated into four half-day modules, covering market, product and technology aspects in turn, before developing the draft roadmap in the final workshop.
- The structure of the roadmap is developed as part of the workshop process, and so does not need to be defined during the planning stage.
- Linkage analysis grids are used as a key part of the process, to explore the relationships between market, product and technology perspectives, and to prioritise efforts.

The T-Plan approach is summarised in Fig. 4, with workshop agendas and activities described in more detail in Table 1.

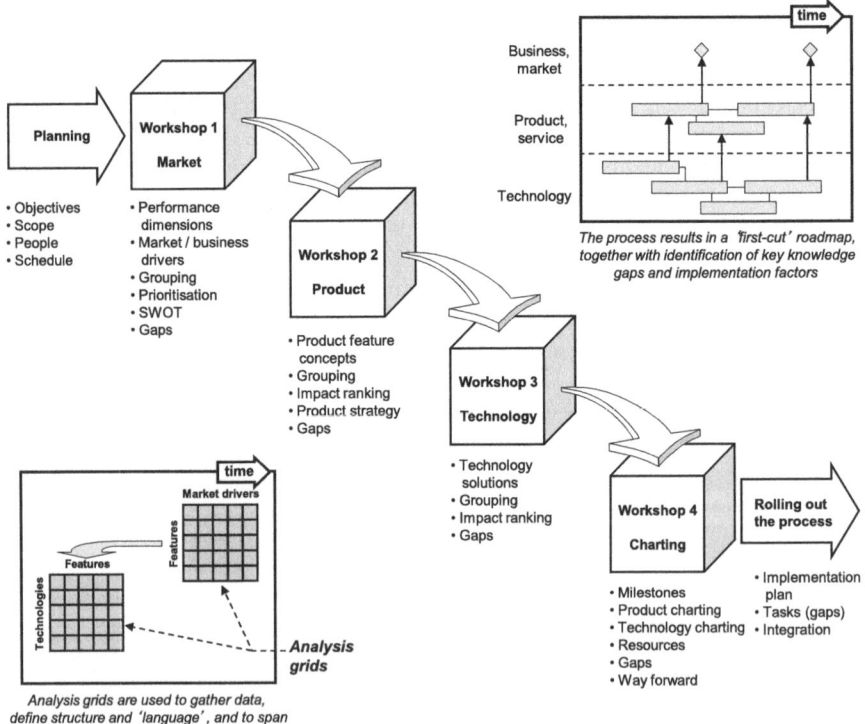

Fig. 4 T-Plan process for product-technology strategic planning

Fast-Start Roadmapping Workshop Approaches

Table 1 T-Plan workshop schedule

	Workshop 1 Market	Workshop 2 Product	Workshop 3 Technology	Workshop 4 Charting
09:00	*Introduction* Workshops seldom start on time, and as a rule of thumb ten minutes should be allowed for this initial delay. Typically the background and aims should be presented by the organisational sponsor, handing over to the facilitation team to run through the agenda and approach, allowing time for a round of brief introductions from participants.	*Introduction* Outputs from the first workshop should be summarised briefly, and participants reminded of the aims, approach and agenda.	*Introduction* Outputs from the first two workshops should be summarised briefly, and participants reminded of the aims, approach and agenda.	*Introduction* Outputs from the first three workshops should be summarised briefly, and participants reminded of the aims, approach and agenda.
09:15		*Product feature concepts* Based on the market (customer) drivers identified in Workshop 1, as many product features, functions and performance aspects as possible should be brainstormed, including service opportunities if appropriate. It is important to avoid discussion about technology at this stage – focus on 'the product brochure of the future'. These should be grouped into clusters to define areas for potential innovation (typically ten or less), so that their potential impact on the market drivers can be assessed in the next session. These product areas are used to define sub-layers in the roadmap (Workshop 4).	*Technological solutions* Based on the product feature concepts, functions and performance requirements identified in Workshop 2, as many technology solutions as possible should be brainstormed. These should be grouped into clusters to define areas of technical capability (typically ten or less), so that their potential impact on the product drivers can be assessed in the next session. These technology areas are used to define sub-layers in the roadmap (Workshop 4).	*Roadmapping* Confirm the focus, scope and format of roadmap, corresponding to a large prepared template on a wall chart. The product and technology sub-layers are defined by the clustered areas identified in Workshops 2 and 3, with the timeframe sufficient for depicting 2-3 innovation cycles. The product vision and strategic requirements are established, by broadly specifying the product functionality and performance for each version. This requires a negotiation between the market / commercial and technology / resource perspectives, to balance the pull and push tradeoffs and alignment. The market, product and technology impact assessments from Workshops 1-3 provide a rationale for guiding this discussion.
09:30	*Performance dimensions* Although the focus of the first workshop is on the market, and in particular customer purchase motivations, it has been found that tackling this directly can be challenging, and a quick brainstorm of key product performance dimensions provides a good starting point (for example, ease of use, cost and size). The most important performance dimensions should be highlighted.			
10:00	*Market / business drivers* External market drivers are brainstormed (customer purchase motivations), using the performance drivers from the previous session as a starting point ('a level up', addressing the question 'why are these performance dimensions valuable?'). These are grouped into clusters (typically ten or less). The process is repeated for internal business drivers, to understand other strategic factors that influence the innovation choices available to the company.			
10:30				
10:45		*Break* The break provides an opportunity for the facilitation team and company sponsor to review progress and deal with any emerging issues.		*Break* The break provides an opportunity for the facilitation team and company sponsor to review progress and deal with any emerging issues. *Roadmapping continued...* Technical programmes are defined, and other aspects of the innovation strategy discussed, guided by the roadmap structure (for example, market drivers, competitors, customers, business strategy, services, operations and resources). Key linkages are mapped, decision points identified and risks reviewed.

Table 1 (*continued*)

	Workshop 1 Market	Workshop 2 Product	Workshop 3 Technology	Workshop 4 Charting
11:00	*Break* The break provides an opportunity for the facilitation team and company sponsor to review progress and deal with any emerging issues.		*Break* The break provides an opportunity for the facilitation team and company sponsor to review progress and deal with any emerging issues.	
11:45		*Impact of product features* A linkage analysis grid is used for this purpose, as illustrated in Fig. 6 and 7. The market / business drivers and product areas define the columns and rows, respectively. Each row is taken in turn, and the question asked 'if we were to invest in this area, which drivers would be satisfied?' (a 'push' question), with the relative impact of each product area ranked for each market and business driver. Then the grid is 'balanced' by reviewing each column, and asking the question 'if we want to satisfy this driver, which product areas have the most potential?' (a 'pull' question), and the scores are adjusted as appropriate. The overall score for each product area is then calculated, weighted by the relative importance of the market drivers.		
12:00	*SWOT analysis* The strategic context within which the product innovation will take place is reviewed, using a SWOT framework to brainstorm external opportunities and threats, together with internal strengths and weaknesses.	*Product strategy* If time allows, the product strategy is reviewed to understand the overall innovation approach to be followed. For example: what are the key differentiators? what price point is appropriate? what product platforms will be developed? what might the product family look like?		
12:15	*Gaps* As a final step, key market knowledge gaps are identified (for example, customers, competitors and legislation), some of which may be addressed before the second workshop.	*Gaps* As a final step, key product knowledge gaps are identified (for example, customer requirements and competitor positions), some of which may be addressed before the third workshop.	*Gaps* As a final step, key technology knowledge gaps are identified (for example, state of the art, technical forecasts), some of which may be addressed before the final workshop.	*Gaps / way forward* As a final step, review key learning points and actions, relating to the innovation strategy, and also the roadmapping process itself.
12:30	*Close* The outputs from the workshop should be transcribed and circulated to participants, and any actions arising from the workshop undertaken.	*Close* The outputs from the workshop should be transcribed and circulated to participants, and any actions arising from the workshop undertaken.	*Close* The outputs from the workshop should be transcribed and circulated to participants, and any actions arising from the workshop undertaken.	*Close* The outputs from the workshop should be transcribed and circulated to participants, ensuring that actions are undertaken and that the process is taken forward as appropriate.

3.2 T-Plan Case Example – New Product Development (Software)

A small software company based in the UK was considering a major re-development of an older software product, aimed at a new niche market in the pharmaceuticals industry (Phaal *et al*, 2001). Technology roadmapping provided a means for supporting product planning, to assess the viability of the proposed development.

The standard T-Plan process was followed, with four half-day workshops attended by about 12 senior members of staff from both technical and commercial functions. Figure 5 shows the main outputs from the second (product) workshop – the market-product linkage grid. This summarises the customer and company drivers and the main product areas, and the relationships between these. A weighted scoring system enables product areas to be prioritised on the basis of their contribution to the market and business drivers. A similar grid is used to link product to technology areas.

Figure 6 shows a transcription of the outputs from the final workshop, where the commercial and technical views are brought together to define the product strategy in detail, together with the associated technology developments. A key part of the debate within the workshop related to the timing of features in terms of software version release dates. This revolved around the tension between the desirability of a feature (market pull) and the capability and effort required to develop the feature (technology push). The outputs from the first three workshops provide a basis for addressing this issue, to help prioritise feature development, and achieve an alignment between market, product and technology strategy.

On the basis of this roadmapping activity it was decided not to proceed with the product development, because the required investment in staff and facilities was

Product Feature Concepts	Market / Business Drivers	Customer					Business			Prioritisation	
	Prioritisation:	9	10	7	...	6	5	5	...	6	
		1. Time to market of drug	2. Integrity of trial	3. Cost of trial	:	7. Ease of use	8. Future proof	A. Reusability	:	D. Motivation of staff	Customer	Softco
1. Security			////	//				/			5.5	4.5
2. Validated software			///	/			/	/		//	4.7	1.1
3. Compliance		//	////	//		//				X	8.7	4.9
...											...	...
10. User friendly		/	/	/		///		/		//	3.2	5.2

Fig. 5 Market-Product linkage grid (extract)

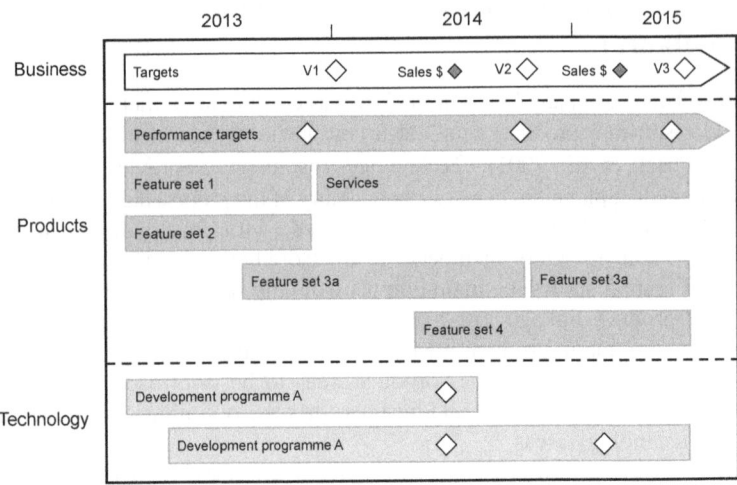

Fig. 6 First-cut roadmap (schematic)

too great a step change for the company given its size. This was considered to be a positive result, avoiding the large waste of effort and management attention that might have resulted if they had proceeded. The company subsequently applied the roadmapping approach again within another part to their business, to support the development of a core product.

4 Summary

Fast-start workshops provide an efficient means of using roadmapping thinking to facilitate strategic dialogue. Wall charts provide structure, used as templates to guide activities, capture perspectives, explore concepts and summarise outputs (see again Fig. 1). Participants are empowered to contribute and interact through the use of sticky notes within facilitated group work sessions. The process is tolerant to diversity, enabling rapid progress in complex business and organisational contexts.

The S-Plan and T-Plan methods support initiation of roadmapping at strategic and product-technology levels, and can also be used as one-off problem solving tools. Fast-start workshop approaches are designed to be agile, in that they are flexible, rapid and problem-focused, suitable for application in both small and large organisations. A light-touch approach to roadmapping can provide an efficient and effective solution, aimed at strategic concerns of immediate relevance and encouraging a focus on the most important issues.

The process learning from the first pilot and subsequent applications can be used to plan the way forward, in terms of both strategy implementation and the roadmapping process itself.

References

De Laat, B., McKibbin, S.: The effectiveness of technology road mapping – building a strategic vision, Dutch Ministry of Economic Affairs (2003), http://www.ez.nl

Phaal, R., Farrukh, C.J.P., Probert, D.R.: T-Plan: the fast-start to technology roadmapping - planning your route to success. Institute for Manufacturing, University of Cambridge (2001)

Phaal, R., Farrukh, C.J.P., Probert, D.R.: Strategic roadmapping: a workshop-based approach for identifying and exploring innovation issues and opportunities. Engineering Management Journal 19(1), 16–24 (2007)

Authors

Robert Phaal is a Principal Research Associate in the Engineering Department of the University of Cambridge, based in the Centre for Technology Management, Institute for Manufacturing. He conducts research in the area of strategic technology management, with particular interests in technology roadmapping and evaluation, emergence of technology-based industry and the development of practical management tools. Rob has a mechanical engineering background, with a PhD in computational mechanics, with industrial experience in technical consulting, contract research and software development.

Clare Farrukh is a Senior Research Associate in the Engineering Department of the University of Cambridge, based in the Centre for Technology Management, Institute for Manufacturing. Research interests include strategic technology management tools, new product introduction, technology valuation and industrial emergence. Clare has a chemical engineering background, with industrial experience in process plant and composites manufacturing, involving engineering projects, production support, process improvement and new product introduction work.

David Probert is a Reader in Technology Management and the Director of the Centre for Technology Management at the Engineering Department of the University of Cambridge. His current research interests are technology and innovation strategy, technology management processes, industrial sustainability and make or buy, technology acquisition and software sourcing. David pursued an industrial career with in the food, clothing and electronics sectors for 18 years before returning to Cambridge in 1991.

Technological Overall Concepts for Future-Oriented Roadmapping

Günther Schuh, Hedi Wemhöner, and Simon Orilski

Rapid technology growth and globalization have led to an increase of complexity of technology management. The roadmapping process supports technology management by providing a way to identify, evaluate and select strategic alternatives to achieve a specific technological objective. The ensuing dilemma is that the development of a roadmap requires information about future events which at an early stage may be questionable. In order to resolve this dilemma, the semiconductor industry for example uses a technological overall concept, referred to as Moore's Law. This technological overall concept serves as a Self-fulfilling Prophecy and is considered to provide orientation in the roadmapping process. This chapter refers to two different types of technological overall concepts – a sector-wide and an enterprise-specific one. The objective of this chapter is to describe the benefits, drivers, development processes, applications and successful practices of these concepts with a view to making them usable in different types of enterprises.

1 Challenges for Future-Oriented Roadmapping

The competitiveness of technology-oriented enterprises depends upon technologies. In order to achieve sustainable success, enterprises cannot allow themselves to be taken by surprise as concerns technological changes and market developments. This appears to be particularly difficult in times of rapid technological development and changes, globalized markets and increasing technological

---

Günther Schuh
Werkzeugmaschinenlabor WZL,
der Rheinisch-Westfälischen Technischen Hochschule Aachen, Manfred-Weck Haus,
Steinbachstraße 19, 52074 Aachen, Germany
e-mail: G.Schuh@wzl.rwth-aachen.de

Hedi Wemhöner
Fraunhofer-Institut für Produktionstechnologie IPT, Technologiemanagement,
Steinbachstraße 17, 52074 Aachen, Germany
e-mail: hedi.wemhoener@ipt.fraunhofer.de

Simon Orilski
GNS Gesellschaft für Nuklear-Service mbH, Frohnhauser Straße 67,
45127 Essen, Germany
e-mail: simon.orilski@rwth-aachen.de

complexity. Hence, the early detection of medium to long-term trends and developments which enable an identification, evaluation and derivation of concrete technological opportunities and potentials represents a major challenge. The development and introduction of new technologies can be regarded as long-term strategic tasks (Gassmann and Sutter, 2008). Therefore they have to be in line with the overall business strategy. Consequently, the inclusion of technological considerations in strategy and planning activities becomes an even greater necessity.

Roadmapping is a process that aims at identifying, evaluating and selecting strategic alternatives which can be utilized to achieve a desired technological objective. The respective challenge is rooted in the fact that – while the roadmap is being developed – the strategy has to be aligned with potential future events. Furthermore, relevant boundary conditions may change. Due to the long-term direction of planning, the reliability of the information on which decisions are based is bound to be questionable; so the longer the range of planning has to be, the less reliable the information that influences decisions becomes. Which means that while the time horizon expands, planning security decreases. Concrete planning is no longer possible. The semiconductor industry for example, resolves this dilemma by use of a technological overall concept known as *Moore's Law*. *Moore's Law* refers to a *Self-fulfilling Prophecy* which provides orientation for market cycles for new products as well as new production technologies within the industry. The technological overall concept is considered to provide long-term orientation in the roadmapping process. The individual enterprise's activities and strategies can respectively be aligned with these demands. This chapter means to discuss the development and use of such technological overall concepts within an entire industrial branch as well as in specific enterprises.

2 Deployment of Technological Overall Concepts

Roadmapping involves the integration of two different objectives: the forecasting of future developments and the planning of own approaches and actions (Kappel, 2001). This brings up the problem that strategic decisions depend on prospective outlooks which are based on present-time knowledge. Thus, the reliability of this information is anything but guaranteed in most cases.

Related to a short or medium-term perspective, concrete information, such as established market and customer requirements, can be used to define the demands a company's technology strategy has to meet. In contrast, the requirements which a technology strategy has to conform to in the long run are difficult to identify and may change dramatically in the course of time. As illustrated in figure 1 by means of the example of market demands, there is a certain point in time at which the planned strategic and technological flexibility does not suffice to grant a successful reaction to changes in market needs.

The described uncertainty of long-term information does not only apply to market demands, but to practically all information that is relevant to the technology planning process – such as future technologies and the prospective scope of products offered by the company. This uncertainty gives rise to the question how a long-term orientation for the technology planning process may be attained.

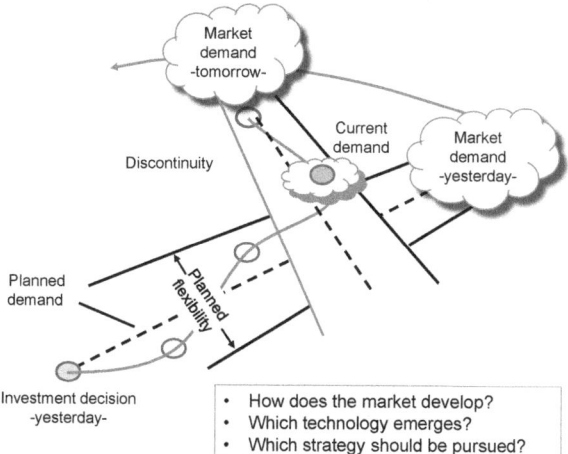

Fig. 1 Market uncertainties within the roadmapping process

The impact of uncertain long-term information and thus the necessity of long-term orientation within the technology planning process can be seen in the development of video systems. This example from industrial history proves that objectively superior technologies are not necessarily more likely to penetrate the market. In the market the VHS system prevailed over the video system Betamax, although in direct comparison, the VHS system was technologically inferior. The success of a technology may thus not only be based on objective advantages but on the definition of a *Self-fulfilling Prophecy* within the industry. This means that all market participants can have a strong influence on the penetration of a technology if they invest in the same direction. Generally each party invests in a research area that strategically suits its enterprise best. When market demands change as depicted in figure 1, enterprises have to make significant investments to *correct the adopted course* (see figure 1). If all enterprises of a particular industry agreed on a coordinated approach – as in the given example from the semiconductor industry – and formulated an overall technological concept, they could significantly heighten the effectiveness of their technological and strategic decisions. By means of devising technological overall concepts, enterprises are able to focus on a common target or a consistent strategy and thus achieve an optimally coordinated performance.

Technological overall concepts are like landmarks to enterprises as they help to make strategic long-term decisions in line with the overall business and technology strategy. Since technology roadmaps represent the operationalization of a technology strategy, technological overall concepts support the roadmapping process. In fact, technological overall concepts are apt to compensate for missing and uncertain requirements in long-term technological forecasts. Furthermore, technological overall concepts provide orientation to evaluate technological options in the context of strategic technological objectives. The chief advantage of technological overall concepts is their ability to give enterprises a direction without restricting their innovation processes (Koolmann, 1992). In the following chapters, two different forms of technological overall concepts will be discussed. In the first part, *sector-wide technological overall concepts* will be explained, while the second part will focus on *enterprise-specific technological overall concepts*.

3 Sector-Wide Technological Overall Concepts

Sector-wide technological overall concepts are characterized by being formulated and pursued by a large number of enterprises in a particular field of industry. The semiconductor industry may serve as a prominent example: Based on *Moore's Law*, five different associations from the USA, Europe, Japan, Korea and Taiwan developed the *International Technology Roadmap for Semiconductors (ITRS)* which provides orientation for the coordination of research and development activities in private and public institutions. *Moore's Law* states that the density of a transistor on a chip doubles each 18 months (Intel, 2005). The semiconductor industry has been operating along these guidelines for many years. This implies that if the use of sector-wide technological overall concepts were transferred to other branches, substantial win-win situations could be achieved for all involved parties. The respective enterprises would be in a position to influence the dynamics and objectives of their industry's development. Benefits for individual enterprises and the entire sector that may result from the implementation of sector-wide technological overall concepts will be described in the next chapter. Furthermore, drivers will be taken into account, followed by a practical example from the aerospace industry and a brief review of successful implementations shall round up the presented findings on sector-wide technological overall concepts.

3.1 Benefits of Sector-Wide Technological Overall Concepts

There are three major benefits that can be attributed to sector-wide technological overall concepts:

- Increased effectiveness of technological and strategic decisions
- Coordination of research and development activities
- Alignment of life-cycles across the entire branch including customers and suppliers

Increased effectiveness of technological and strategic decisions: By aligning strategic and technological decisions concerning sector-wide technological overall concepts, enterprises ensure that their planned flexibility is sufficient for meeting future requirements (see figure 1). The reliability of information increases and the effectiveness of strategic decisions increases accordingly.

Coordination of research and development activities: By the implementation of sector-wide technological overall concepts, the research and development activities of all enterprises in a particular sector are oriented towards the same direction.

This enables the realization of a comprehensive technological cooperation as all involved enterprises are guided by the same objectives. Coordinated research and development activities do not only facilitate cost-intensive and difficult projects but also reduce research and development expenses. Moreover, the conjoined approach clears the path for projects that depend on coordination for success, e.g. the setting of technological standards to enable mass production, low production costs and the market success of new technologies, particularly in competitive conditions.

Alignment of life-cycles across the entire branch, including customers and suppliers: Figure 2 illustrates how sector-wide technological overall concepts influence the definition of technology life-cycle phases and the benefits for involved enterprises. It depicts several product life-cycles over a certain period of time. Enterprises are faced with the question, when and for what reason established products should be replaced by new products. The curve of costs (= costs per piece) is outlined as well as the curve of performance (= performance standard of the particular product). The question of how long a product lifecycle lasts depending on costs and performance cannot be answered unequivocally. Technology-oriented branches of industry are marked by the tendency to replace recent products by innovative ones, even though their technical standard still meets current requirements. Consequently, enterprises are forced to improve their latest technologies in functionality and performance in order to maintain competitiveness. This is all they can do to convince consumers of the necessity to replace a recent appliance by a new acquisition, regardless of the fact that the old one still serves its purpose. The enterprises of a branch have to consider what type of technological factor could be utilized for motivating the consumer to buy a product of the new generation, and – consequently - in what kind of performance feature or technological facility they ought to invest. Further questions to be taken into account are: When should those features be available? And what would be an appropriate price?

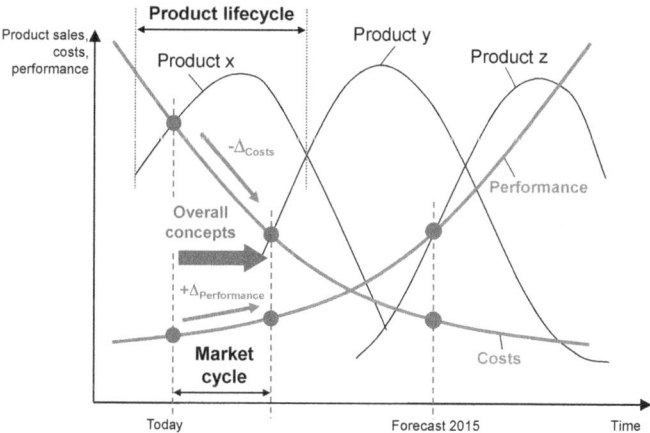

Fig. 2 Effects of sector-wide technological overall concepts

In the example shown in figure 2 a reduction in costs (-ΔCosts) causes the replacement of product x by product y (new product). The product change is thus motivated by financial reasons. However it might be advisable not to base product changes only on economic effects but on significant performance improvements. A specific characteristic of the implementation of sector-wide technological overall concepts is that the technology life-cycles are not defined by individual enterprises. The effect of sector-wide technology overall concepts lies in the definition of costs and performance for a certain period of time. Also, the guidelines of this concept help suppliers and end consumers to orient themselves within a branch. In the semiconductor industry, for instance, the life-cycles of semiconductor manufacturers are aligned with those of customers (e.g. computer manufacturers) as well as those of suppliers (e.g. lithography technology manufacturers) by means of the technological overall concept defined in the *International Technology Roadmap for Semiconductors (ITRS)*.

3.2 Drivers of Sector-Wide Technological Overall Concepts

Sector-wide technological overall concepts can emerge from different drivers. In the following chapter, the main drivers and the different applications resulting from them will be observed:

Technological dependency of the enterprises within a sector: If the enterprises belonging to a particular industrial sector are marked by a high degree of dependency upon one another, the implementation of sector-wide technological overall concepts helps to coordinate life-cycle phases as well as research and development activities. By coordinating these activities, significant technological development can be achieved within the branch. This is of particular importance if the branch comprises a high quantity of small companies which separately are unable to set technological standards or ensure the market penetration of a new technology. Here, technological and market development can only be influenced by joint efforts. For instance, a strategic process for the development of a technological overall concept has been established in the German biotechnology branch (Biotechnologie, 2010).

Consistency of technological evolution: The aim to influence the technological evolution within an industrial sector in order to achieve a consistent long-term course is a typical motivation for the application of sector-wide technological overall concepts. A perfect example of this can be found in the semiconductor industry.

Shared sector-wide interests: This refers to the need for a sector-wide cooperation to promote a certain direction of development or to achieve specific objectives. Said objectives may consist in the fulfilment of legal requirements, the setting of standards or the desire to influence political or social conditions (lobbying). Interests of this kind can only be enforced if the various parties of a sector cooperate and focus on the same direction.

3.3 Development and Application of Sector-Wide Technological Overall Concepts

Subsequently, a practical example from the European aeronautic industry will serve to explain the creation and application of sector-wide technological overall concepts.

In the aeronautic industry a sector-wide technological overall concept was developed by the *Advisory Council for Aeronautics Research in Europe (ACARE)*. In the year 2000, the European Commissioner for Research Philippe Busquin invited the starting group, which consisted of 15 high-ranked experts from the airport, airline and air traffic sectors, regulators and airframe, engine and equipment manufacturers to map out an intermediate to long-term vision of future aeronautics. As the aeronautic industry comprises a large quantity of interdependent companies on various supplier levels, the common objective was to align the research and development activities of all involved players. Through collaboration and guided by a shared vision, Europe was to become the world's number one location for aeronautics.

In 2001, this task group produced the *Vision 2020* - a 26 - paged manuscript, issued for the interest of industrial stakeholders, European policy-makers, national institutions and the broader public alike. This manuscript displays a vision of the European aeronautical sector in the year 2020, with regard to consumers' choice, comfort and costs (Acare, 2001). In order to materialize this vision and its goals, a *Strategic Research Agenda (SRA)* was launched in 2002. Two years later, a more detailed second edition (SRA 2) was elaborated. The group was joined by an ever-growing number of new members, including representatives of various European states, the commission, the group of stakeholders, the manufacturing industry, airlines, airports, service providers, regulators, research establishments and the academic field.

As shown in figure 3 the SRA is divided into five main categories: Quality and affordability, Environment, Efficiency of the air transport system, Safety and Security.

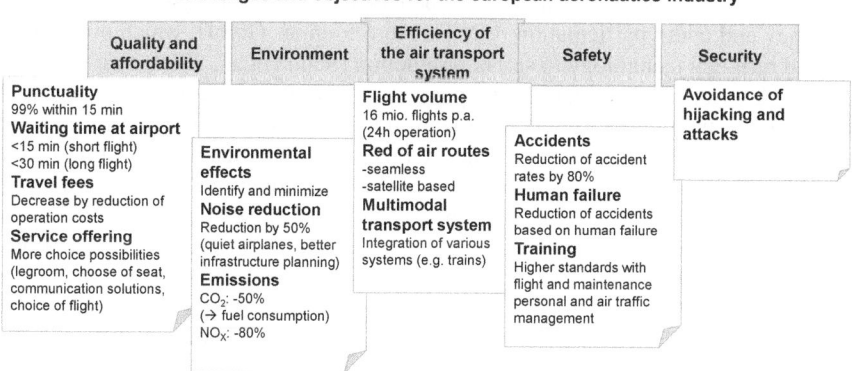

Fig. 3 Organization of the Strategic Research Agenda

Sector: aircraft	Technology cluster: **aerodynamics / combustion / air-breathing propulsion / nozzles, vectored thrust, reheat**		
Flight physics	Base		
Aero structures	Key *Aerodynamics* • Reduced thermomechanical distortions • More efficient sealing technology		
Propulsion			
Aircraft avionics, systems & equipment	*Combustion* • Enhanced mixing design • Technologies for lean combustion • Multi-point fuel injection	*Air-breathing propulsion* • Integrated nacelle/ wing design for UHBR engines • Nacelles technologies for weight reduction	*Nozzles, vectored thrust, reheat* • Thrust reverser, technologies for weight reduction
Flight mechanics			
Integrated design & validation (methods & tools)	Pacing		
Air traffic management		• Geared fan engine • Ultra High Bypass Ratio engines (UHBR engines) • Contra-rotating fan engine	• Provide low noise thrust reversal capability
Airports			
Human factors			
Innovative concepts & scenarios	Emerging 2010	2015	2020

Fig. 4 Contents of technology cluster in the aircraft sector

Each of these five categories of the SRA is in turn sub-divided into three different sectors, namely Aircraft, Airport and Air traffic management (ATM), and ten technology fields, such as propulsion, flight mechanics, human factors and others. Within these sectors, the goals of the SRA are detailed according to concrete technologies, clustered in technology roadmaps. A technology roadmap for the aircraft / propulsion sector is outlined in figure 4.

The aim to reduce CO_2-emissions (in the environmental category) is substantiated by the projected development of a geared fan engine until 2015 (see figure 4) (Acare, 2002 and Acare, 2004). The gear enables a slower rotation of the fan in comparison with a turbine. Thus, it becomes possible to utilize larger fans and attain a higher bypass ratio. The result is a highly energy efficient engine that reduces fuel consumption by 15% and is also characterized by a significantly lower noise-level. Despite their obvious merits, geared turbofans used to be a niche product. In this case, the ACARE and the Strategic Research Agendas instigated a cooperation of several machinery component manufacturers, guided by Pratt & Whitney and under participation of MTU Aero Engines GmbH, which brought the geared turbofan technology to series-production (Schuh et al., 2009).

The Advisory Council (ACARE) possesses a structure of governance in which a plenary council and an integration team review strategic options, priorities and objectives. Supporting groups provide forums for communication. By pointing out the need for research, the ACARE effectuates a joint commitment of different parties within a sector in spite of competition. European as well as national programs and other initiatives also refer to the need for research and initiate ambitious projects. One example is the Clean Sky Initiative. The formation of private consortiums and co-operations without involvement in European or national programs also happens. In the framework of such programs and consortiums, different enterprises have the opportunity to collaborate. The mutual commitment of otherwise

Technological Overall Concepts for Future-Oriented Roadmapping   115

competing companies is based on the fact that the ambitious and complex objectives of the research programs rely on united efforts. However, each involved enterprise can still accomplish individual sub-objectives. Beyond that, the enterprises of the aeronautic sector strive to attain the goals set by the ACARE, as nearly all of them have a representative in the ACARE council. According to the bottom-up principle, the goals of the ACARE are transported into the enterprises. The formation of the ACARE council was, on the one hand, driven by legal regulations, e.g. the reduction of CO_2-emissions, and on the other hand, by the common desire to strengthen European aeronautics (Acare, 2011).

3.4 Successful Implementations

The question arising now is what factors happen to support a successful implementation of sector-wide technological overall concepts. In the following, some of the chief success factors will be summarized.

Innovation space for enterprises: Each enterprise that participates in the process of developing a technological overall concept has to perceive an individual benefit. The attainment of the technological overall concept's objectives must never interfere with the competitiveness of the individual enterprise. Thus, technological overall concepts typically refer to pre-competitive research areas.

Time horizon: The formulation of sector-wide technological overall concepts has to involve a time restriction. *Moore's Law* sets a time frame of 18 months and the ACARE's Vision is *deadlined* to 2020. This time horizon must be suitable for the sector in which the technological overall concept is intended to find application. As regards the aeronautic industry, 20 years seem to be appropriate, according to the length of research, development and manufacturing phases. For the semiconductor industry, 18 months are a feasible time horizon.

Combination of technological and economic aspects: The formulation of the sector-wide technological overall concept should also take economic aspects into account. The consideration of economic aspects is of particular importance as it ensures the commitment of the process participants to defined technological objectives. Furthermore, the economic perspective links the technological overall concept to environmental market conditions.

Stakeholder Commitment: The stakeholders' commitment to the technological overall concept is an indispensable precondition of its success. In order to attain this commitment, the technological overall concept has to effectuate significant economic benefits for the individual stakeholder without endangering competitive advantages. A high acceptance can only be achieved, if the technological overall concept's development is based on the cooperation of all involved stakeholders. Additionally, the overall concept has to be expressed in a concrete, transparent and action guiding form.

Suitable starting groups: As mentioned above, the starting group has to comprise representatives from the most important participating enterprises or institutions. The example of the ACARE Vision shows that a group size of 15 members is suitable for devising a complex concept. The representatives should be of a sufficiently high rank to generate support in their respective enterprises or institutions. Diversity within the team is the basis of objectivity.

Realistic and challenging objectives: The objectives set by the sector-wide technological overall concept need to be ambitious, but realistic. On the one hand, the objectives must be challenging enough to not allow realization by a single company, as this underpins the benefit of the technological overall concept. On the other hand they have to stay feasible and achievable.

Appropriate degree of concreteness: In order to implement the technological overall concept within the sector, the objectives have to be formulated precisely, but without going too far into detail, as that would deprive the individual enterprise of innovation space.

Actuality: The SRAs have to be kept up-to-date. Otherwise, their action guiding potential and acceptance will suffer. Consequently, technological overall concepts require revision on a defined regular basis.

4 Enterprise-Specific Technological Overall Concepts

Contrary to sector-wide technological overall concepts, enterprise-specific technological overall concepts are characterized by being formulated and pursued individually by every enterprise of the sector. They are used to align technology development and technology deployment within the enterprise. This addresses the problem of uncertain information in long-term planning. In the long-term perspective, enterprise-specific technological overall concepts supersede the concrete information that provide landmarks in short- and medium-term planning, e.g. market and customer demands, defined future products or specific technological information. They support the alignment of technology development and technology deployment in long-term planning.

4.1 Benefits and Drivers of Enterprise-Specific Technological Overall Concepts

Enterprise-specific technological overall concepts help to coordinate overall technology deployment with a long-term horizon. If a technology portfolio does not dovetail with the overall concept, the gaps are pointed out and can be filled accordingly. Thus, enterprise-specific technological overall concepts facilitate the identification of *white spots* in the technology portfolio of an enterprise.

Furthermore, they serve to assess technological ideas at an early stage and align them with the strategic orientation of the enterprise. In the automotive industry, for example, the technological overall concept *economic lightweight construction* supports the orientation of technology pre-development (Schuh et al., 2008). Every technological idea is assessed with regard to its contribution to the technological overall concept before pre-development. Having passed the pre-development process, the feasibility of these ideas is assured, so they can undergo further development and find application in future automotive projects. The challenge of technology pre-development in the automotive sector lies in the lacking allocation to actual vehicles. In the early phases of technology pre-development, it is not yet clear which particular technology will be utilized in which particular vehicle. Consequently, there are no concrete requirements discernible. Enterprise-specific technological overall concepts compensate for these missing requirements and thus help to align pre-development with the objectives of the technology strategy.

4.2 Development and Application of Enterprise-Specific Technological Overall Concepts

For the development of an enterprise-specific technological overall concept the following steps have to be considered:

Step 1 - Defining objectives of enterprise-specific technological overall concepts: First of all, the objectives and the purpose of the enterprise-specific technological overall concept have to be defined. In general, enterprise-specific technological overall concepts help aligning technology development and deployment. In detail, enterprise-specific technological overall concepts can, for example, be used to assess the contribution of new technological ideas to a defined technology strategy on the whole, or to coordinate the timing of technological developments. In the following step, the technological overall concept's content depends on its objectives.

Step 2 - Specifying the content and elements of technological overall concepts: The content and the elements of enterprise-specific technological overall concepts have to be specified. This specification depends on the overall concept's objectives. In case of the example from the automotive industry this means, that the technological overall concept *economic lightweight construction* must address the technology clusters and fields to be processed as well as the technological objectives of the enterprise as a whole. This ensures an orientation of pre-development according to the technological objectives of the enterprise and the appropriate technological fields.

Step 3 - Detailing of enterprise-specific technological overall concepts: The basis of the development of an enterprise-specific technological overall concept is formed by the enterprise's technology strategy. As mentioned before, the technology strategy includes significant statements regarding technology fields to be considered, the performance level at which technologies are employed within these fields, the time at which these technologies are to be launched on the market, the source they are obtained from and the possibilities of their profitable application. These statements have to be taken into account while the enterprise-specific technological overall concept is being developed. The terms in which a technological overall concept substantiates the technology strategy have to be specific enough to enable its realization in concrete applications.

Step 4 - Communication and application of enterprise-specific technological overall concepts: The enterprise-specific technological overall concept must be communicated to the relevant business units and divisions of the enterprise. Like the sector-wide technological overall concept it must be kept up-to-date and be adapted to environmental changes on an adequately regular basis.

4.3 Successful Implementations

The question that was asked in the chapter on sector-wide technological overall concepts also applies to their enterprise-specific counterpart, i.e.: What factors support the successful implementation of enterprise-specific technological overall concepts? Again, some of the chief success factors will be summarized in the following section.

Alignment with technology strategy: The enterprise-specific technological overall concept has to be aligned with the enterprise's technology strategy. It must substantiate the most important elements of the strategy.

Operationalization and applicability: The technological overall concept requires concrete description in order to be practically applicable in technology management. Typically, a technological overall concept is more concise than the technology strategy. Also, it does not include all aspects of the technology strategy, but focuses on those that are relevant in regard to the concept's purpose. The enterprise-specific technological overall concept has to be action-guiding and easily comprehensible. Furthermore, it has to facilitate the assessment of technological ideas in terms of their contribution to the realization of the technological overall concept.

Process integration: It is necessary to embed the enterprise-specific technological overall concept in the technology management process. The aforementioned example of the *economic lightweight construction* concept for technology pre-development in the automotive industry provides a guideline for the assessment of technological ideas, which may be transferrable to technology pre-development. Thus, technological ideas can be assessed concretely with regard to their contribution to the realization of the *economic lightweight construction*.

Liability: The enterprise-specific technological overall concept has to be adhered to at all times. For instance, this includes a consistent utilization thereof in the roadmapping process.

Stakeholder commitment: As is the case of sector-wide technological overall concepts, relevant stakeholders have to get involved in the development of the enterprise-specific technological overall concept. This helps to promote the acceptance as well as the future utilization of the concept. Accordingly, management, production, research and development, marketing etc. have to be taken into account.

5 Conclusions

Using the roadmapping method as a technology management tool enables enterprises to systematically identify and evaluate product, material and production technologies, and to align them with their overall strategies. For enterprises, roadmapping is a prerequisite of maintaining competitive capability within the sector. In order to apply the roadmapping method successfully, enterprises require the velocity, agility and flexibility that enables them to identify and utilize market opportunities at an early stage.

The inherent dilemma of the roadmapping process results from the fact that investments as well as strategic decisions rely on prospective outlooks that are based on present-time knowledge. Hence, the reliability of this information can hardly be taken for granted.

Technological overall concepts support a flexible adaptability to prospective technology requirements and market needs. Furthermore, they significantly increase the capacity to identify future technologies at an early stage. Sector-wide technological overall concepts coordinate the research and development activities of an entire sector and thus help to accomplish complex and ambitious project objectives. Enterprise-specific technological overall concepts substantiate and operationalize the technology strategy of an enterprise and help to align the enterprise's technology development and technology deployment.

Accordingly, enterprises should not only generate their own enterprise-specific technological overall concepts but also contribute to the implementation of sector-wide technological overall concepts. In any case, the successful creation of a technological overall concept involves the observance of various influential factors.

References

Acare, A Vision for 2020 (2001), http://www.acare4europe.com/html/documentation.asp (accessed June 28, 2011)
Acare, Strategic Research Agenda 1 (2002), http://www.acare4europe.com/html/documentation.asp (accessed June 28, 2011)
Acare, Strategic Research Agenda 2 (2004), http://www.acare4europe.com/html/documentation.asp (accessed June 28, 2011)

Acare, Advisory Council for Aeronautics Research in Europe (2011),
   http://www.acare4europe.com (accessed June 28, 2011)
Biotechnologie, Auftakt zum Strategieprozess: Ideen zur Biotechnologie der Zukunft gefragt (2010), http://www.biotechnologie.de/BIO/Navigation/DE/root,did=113362.html?listBlId=74462& (accessed June 28, 2011)
Gassmann, O., Sutter, P.: Innovationsprozesse. In: Gassmann, O., Sutter, P. (eds.) Praxiswissen Innovationsmanagement. Von der Idee zum Markterfolg, Hanser, München (2008)
Intel, Excerpts from a Conversation with Gordon Moore: Moore's Law (2005)
Kappel, T.: Perspectives on Roadmaps: How organizations talk about the future. Journal of Product Innovation Management 18, 39–50 (2001)
Koolmann, S.: Leitbilder der Technikentwicklung. Das Beispiel des Automobils. Campus Verlag (1992)
Schuh, G., et al.: Roadmapping: Geschäftserfolg durch zielgerichtete Technologieentwicklung. In: Brecher, C., et al. (eds.) Wettbewerbsfaktor Produktionstechnik. Aachener Perspektiven, Apprimus, Aachen (2008)
Schuh, G., et al.: Technologie-Roadmapping: Erfolgreiche Umsetzung in der industriellen Praxis. Zeitschrift für Wirtschaftlichen Fabrikbetrieb (104), 291–299 (2009)

Authors

Günther Schuh is Professor for Production Engineering at the RWTH Aachen University and one of the directors of the Laboratory for Machine Tools and Production Engineering (WZL) and the Fraunhofer Institute for Production Technology IPT. Furthermore, he has been director of the Research Institute for Industrial Management (FIR) of the RWTH Aachen since 2004. He is founder and proprietor of Schuh & Co. and member of several supervisory boards. From 2008 to 2012 he was prorector for industry and economy at the RWTH Aachen. In 1991 he was awarded with the "Otto-Kienzle"-medal of WGP (German Academic Society of Production Engineering) for merits in interdisciplinary research. His area of work comprises production management, technology management, innovation management and complexity management as well as enterprise networks, production controlling and process optimization.

Hedi Wemhöner studied mechanical engineering and business administration at the RWTH Aachen University. Since 2010 she works as research assistant within the department of technology management at the Fraunhofer Institute for Production Technology IPT. Her area of work is technology planning.

Simon Orilski studied mechanical engineering and business administration at the RWTH Aachen University. From 2005 till 2011 he worked as research assistant within the department of technology management at the Fraunhofer Institute for Production Technology IPT, latest as group manager for technology forecasting. His research focused on technology roadmapping and technology forecasting. Since 2011, he works at the GNS Gesellschaft für Nuklear-Service mbH, currently as project manager.

Scenario-Based Exploratory Technology Roadmaps – A Method for the Exploration of Technical Trends

Horst Geschka and Heiko Hahnenwald

Scenario-based exploratory technology roadmaps are a profound and comprehensive basis for the concrete planning of technologies and innovations. In comparison to other approaches of technology roadmapping a particular focus is put on an intensive analysis of the technology's non-technical influences. This concept assumes that the development of a technology is influenced by market-related, societal and economic factors as well as by technical factors outside the technology under consideration. Another characteristic of this approach is that the development of the technology does not follow an evolutionary path. The future development of the impact area is established in the first place. From this future picture of exogenous influencing factors, the performance requirements applying to the technology and the pathways of technological development are derived. Scenario-based technology roadmaps are an instrument of technological forecasting; they are not yet a planning instrument.

1 What Are Scenario-Based Exploratory Technology Roadmaps

In contrast to other approaches to technological forecasting and technology roadmaps, the investigation of the non-technical fields of influence plays a major role in the approach of scenario-based roadmaps. This is based on the thesis that the development of a technology in principle is on the one hand subject to certain inner dynamics, however on the other hand these dynamics are significantly influenced by exogenous factors (e.g. social trends, market trends, laws). Thus, the description of the future development of the technological environment forms a corridor in which the technology roadmap unfolds. The development status of a technology is therefore a result of the exogenous framework and the inner dynamics of the technological development (see Figure 1).

Horst Geschka · Heiko Hahnenwald
Geschka & Partner Unternehmensberatung, Guerickeweg 5,
64291 Darmstadt, Germany
e-mail: {hg,hh}@geschka.de

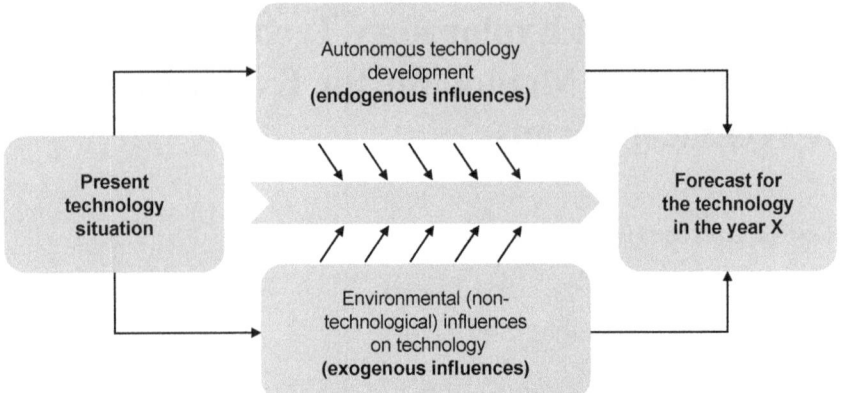

Fig. 1 The non-technical impact area of the technology under investigation

In this connection the development of the technology does not follow an independent evolutionary path. An evolutionary development is often directed by intermediary results and short term views, i.e. without orientation. But a long range perspective cannot turn up through short term steps; this approach is therefore misleading. For this reason, scenario-based technology roadmapping starts with a comprehensive depiction of the influencing factors in the envisaged future year. Then the technological developments are laid down under these external conditions.

Exploratory technology roadmaps are an instrument of technology forecasting. They show how a technology may evolve. They do not contain concrete planning elements, but they supply a solid basis for in-house technology and innovation planning. By comparing the company's existing technology skills and plans to the technology pathways projected in the roadmap, deficits in know-how and resources as well as improvement potentials in terms of the technology could be identified.

2 Basics of Technological Forecasting

Subsequently, we present a method to elaborate scenario-based technology roadmaps as instruments for technology forecasting. They show how a technology may evolve. They do not contain concrete planning elements, but they supply a robust basis for in-house technology and innovation planning.

2.1 Delimitation and Definition of the Technology Field

Above of all, technology forecasting requires a clear delimitation and precise description of the respective topic. A technology is best described by means of three parameters:

- Description of the basic principles: outlining the technical principles and the functional interrelations of structures and processes in a form that is adequate for the specific discipline (technical delineation, proof of function, chemical formula, process diagram, etc.).

- Technical performance indicators: e.g. efficiency, scrap rates, throughput, velocities, fuel or electricity consumption, density, quality characteristics.
- Diffusion characteristics: e.g. number of products in use (absolute figure), number per capita, per household, per companies; percentage of new products based on a new technology of the total market.

The technology of a product or system is embedded in a technological environment. Only this integration guarantees that this technology can be produced, the end-product works without failures and fulfills its final economic purpose (for the following: Geschka, 1994a). If the focus is set on product technology, the following elements of a technological system can be differentiated (see Figure 2):

- Upstream technologies are incorporated into the product, in particular commodities, materials, accessories and components.
- Complimentary technologies are used jointly with the product, i.e. paints and brushes, engine and fuel, camera and film, hardware and software.
- Production technologies influence performance indicators, the styling of the product and production costs.
- Downstream technologies are systems in which the product is integrated or in which it is used, i.e. a navigation system built into a car compatibly, or a test device integrated into a production process. This means that the product has to fulfill certain requirements in order to be compatible.
- Substitutive technologies fulfill basic requirements by a completely different technology; a substitution process takes place. Examples: CDs replace vinyl records, adhesives replace welding, email substitutes surface mail.

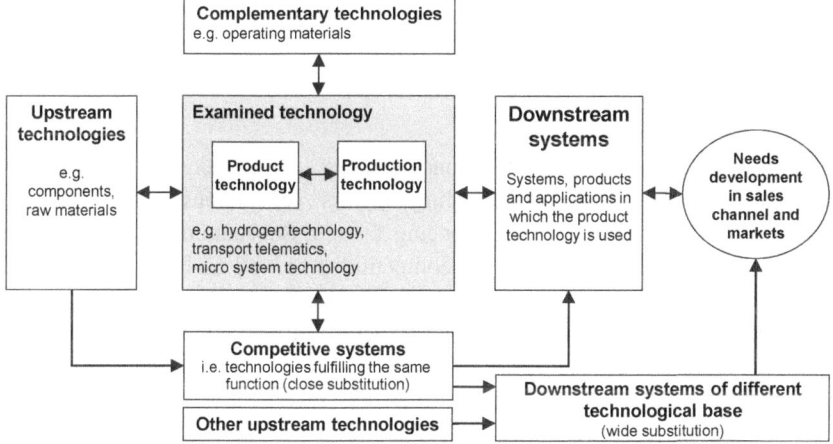

Fig. 2 The technical impact area of the target technology - The technology complex

The technology of a product or a system is incorporated in a relationship of influences ranging from various sub-technologies to process- resp. production-technologies required for manufacturing. This relationship cluster is called technology complex. In the specific case of technological forecasting, the field of investigation has to be delimited clearly and carefully. Its characterization depends on the intent of the forecast as well as on technological circumstances. On the one hand, it might make sense to consider the technology of the product and that of its production simultaneously. For instance, an increase in performance could depend on higher purities of raw materials, which in turn depend on the production process. On the other hand, however, it might be preferential to include the manufacturing of raw materials, as the ecology and the design of the application technology depend on it.

2.2 Knowledge Base of Experts

For the process of technological foresight knowledge of experts is required. Experts possess comprehensive knowledge of the actual situation and the developments and plans for the future in their particular working field. The involvement of experts is of essential importance for the delineation of the topic, the description of the present situation and for the detection, interpretation and projection of future developments.

How exactly expert knowledge can be incorporated, depends on the study's design. There are various options of involving experts in prognostic studies (Geschka, 1994a): e.g. by participation of one crucial expert; by interviewing several experts independently; by getting a group of experts to interact with one another in a workshop; by the anonymous participation and communication between a number of experts in a Delphi study; or, possibly, by combining any of the aforementioned approaches.

2.3 Environmental Analysis

Many technologies possess certain inner dynamics; in addition they are strongly affected by non-technical influencing factors (exogenous influences). We distinguish direct and indirect influencing factors. Substantial direct influencing factors are market factors (demand, competition) as well as laws and standards, which have an immediate impact on a product or a production process. Indirect influencing factors are basic trends in society, politics and economy. Figure 3 shows a general concept of direct and indirect non-technical influences; however, the specific selection of relevant influencing factors always depends on the technology under investigation.

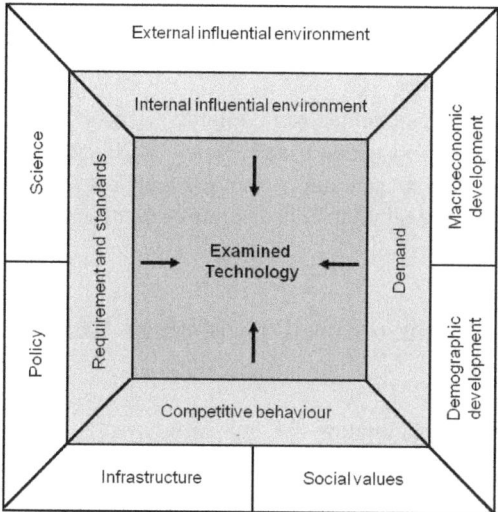

Fig. 3 Input factors for a technology forecast

For the analysis of the external impact area of a certain topic, the scenario technique is well established (Geschka, 1999). By means of this method, predictions of the impact area's future are developed.

The process of the scenario technique consists of three main steps:

1. Identification of exogenous influencing factors on the investigated topic;
2. Projections of the most relevant influencing factors along with arguments for the projection; alternative projections are possible.
3. The alternative projections have to be combined in such a way as to form consistent bundles of projections (scenarios).

The bundling of the alternative projections is achieved with the aid of a so-called consistency matrix: All projections are confronted with one another and it is assessed whether they are fitting well together or are contradictory; this is done according to a scale ranging from "+3" (fits perfectly, is a must) to "-3" (does not fit at all, completely contradictory). The scale value "0" is chosen, if the alternative projections are regarded as coexistent or extraneous to each other. Influence factors that have only one projection are included in this step, because they might influence the selection of alternative projections in other descriptors. Using an algorithm[1], those combinations of projections marked with a particularly high consistency are selected. By way of this procedure a number

---

[1] E.g. The scenario software INKA3, developed by Geschka & Partner Unternehmensberatung, was designed for the development of scenarios and based on the consistency approach.

of alternative scenarios of the topic's influencing area are created. These scenarios of the external impact area - usually only two or three are selected - describe the future situation of the topic.

Applying the scenario technique is a complex and work-intensive process. One of the reasons for this effort is that usually between 20 and 30 descriptors have to be elaborated; in order to generate differentiated prospects of the future it is recommended to describe alternative projections - if substantiated - for most of the influencing factors.

3 Developing Scenario Based Technology Roadmaps (with Examples)

The process for the elaboration of scenario-based exploratory technology-roadmaps consists of three major stages (see Figure 4): Stage 1 and 2 refer to the basic principles, stage 3 refers to the development of the technology pathways.

Stage 1: Identification of the technology field

The starting point of the roadmapping process is a description of the state of the art of the topic technology as well as of all identified elements of the technology complex (see chapter 2.1) (upstream and downstream technologies, complementary and substitutive technologies, production technologies). The actual state of knowledge is examined, especially the state of research and development, patents, pilot products

	Stage 1: Technology analysis	Stage 2: Environment analysis	Stage 3: Roadmap development
Definition	• Definition of the technology under investigation • Identification of all part- and process technologies	• Identification and structuring of relevant non-technical influencing factors on the technology under investigation	• Identification of requirements on the investigated technology and all other elements of the technology complex aligned on the scenario pathways
Description	• Description of the present situation of all technologies in the technology complex	• Description of the present situation and projections on the possible future development of all influencing factors	
Analysis	• Impact analysis of the elements of the technology complex	• Impact analysis of exogenous factors • Bundling and selection of consistent future scenarios	• Consequences for the technology development and interdependencies between the elements of the technology complex
Result	• Description and analysis of the state of the art of the technology complex	• Description of scenario pathways of the technology environment	• Elaboration and visualisation of technology pathways

Fig. 4 Process of scenario-based technology roadmapping

and products that are already on the market, etc. Additionally, all established limitations and barriers regarding the various elements of the technology complex are highlighted. If different materials and commodities in an upstream technology are known to exist and possess potential, all alternatives and their respective pros and cons have to be explored and described. Other limitations, such as legal regulations (e.g. emission rate limits), physical and technical limits and cost limitations (e.g. expensive materials) also have to be taken into account.

The description of the current situation is crucial for an exact definition of the technology under investigation. It has to include the distinction of the product technology from other technologies belonging to the identified technology complex as well as the required structure of the technology in the sub-technologies. The delimitation may require the structuring of the technology into sub-technologies on equal levels. Figure 5 shows the delimitation of the topic (marked grey) for the photo-voltaic technology complex.

However, it is not always recommendable to divide the topic into sub-technologies. Technological differences and, in consequence, different application areas suggest that the roadmapping process should be limited to a close field of application. A delimitation of one significant sub-technology is often more informative than a parallel study of diverging sub-technologies.

The description of the technology complex should also cover the interdependencies between the various technology components. This is achieved by means of a so-called impact matrix. In the impact matrix, the different technologies are compared with each other, and impacts (conducive or impedimental) and interdependencies are determined.

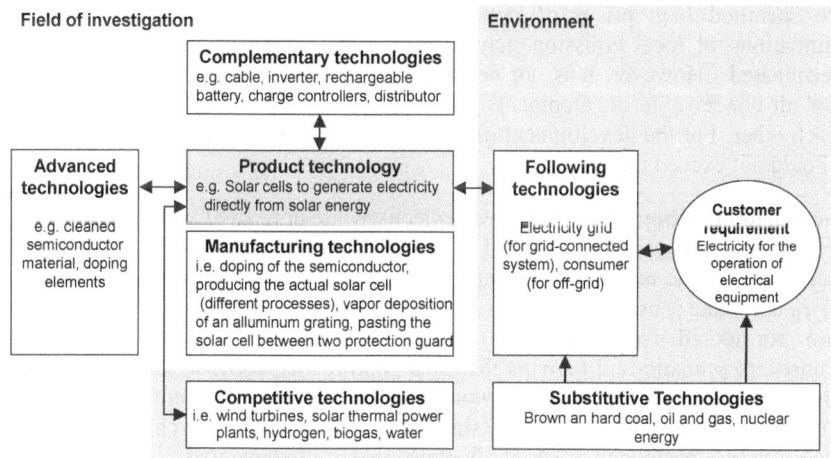

Fig. 5 Technology complex for the photo-voltaic technology

Stage 2: Scenarios for the influential environment

In addition to the impact area that has been identified for the examined technology complex, the non-technical impact area should also be analyzed by determining and describing the impacting factors. This process-stage aims at developing an accurate picture of all interdependencies within the impact area, in order to enable a deduction of consequences for the technology.

Exogenous influencing factors: The list of influencing factors contains some elements which have an effect on the product technology or sub-technologies within the technology complex. These include legal regulations, demographic development, purchasing power, availability of resources, market trends and changes in the competition pattern, as well as the requirements of users and consumers. It is important to select only those factors that have a direct impact on the technology (or the technology complex). The more indirect the impact of the factors is, the more difficult it becomes later to deduce consequences and requirements for the technology. Examples of such non-technical (i.e. technology-specific) influencing factors in the field of fuel cells include: the state of development of drive technologies, user preferences regarding mobility or limitations of local emissions for traffic (see Figure 6). Examples from photovoltaic field are: the law of reimbursement for renewable energy, the cost of non-renewable energy production or the efficiency of photovoltaic in comparison to other renewable energies.

Describing the state of the art: Once the relevant impact factors have been defined, the next step is to describe the state of the art and to develop projections into the future. It is crucial to base these projections on given reasons. If the development of an impacting factor is uncertain, possible future situations should be specified (e.g. prices of raw materials may develop differently, laws on limitations of local emission may change, subsidy grants may be extended or terminated.) However, it is not necessary to formulate and delineate projections for all conceivable developments - only for those that differ significantly from each other. For the development of scenarios the number of projections per factor should not exceed three.

Scenario building: The alternative projections are compared with each other in a consistency analysis and grouped into consistent sets. Theses bundles form the scenarios for the non-technical impact area. For the development of a roadmap it does not make sense to analyze several scenarios. In general, one or two scenarios are considered (see Figure 7a). It is recommendable to select the most consistent scenario and perhaps the most "optimistic" scenario in order to depict the decisions and measures that would lead to a desirable future situation. As a "counter scenario" the most pessimistic scenario could be chosen, in which decisions are postponed or certain features are not implemented.

Scenario-Based Exploratory Technology Roadmaps

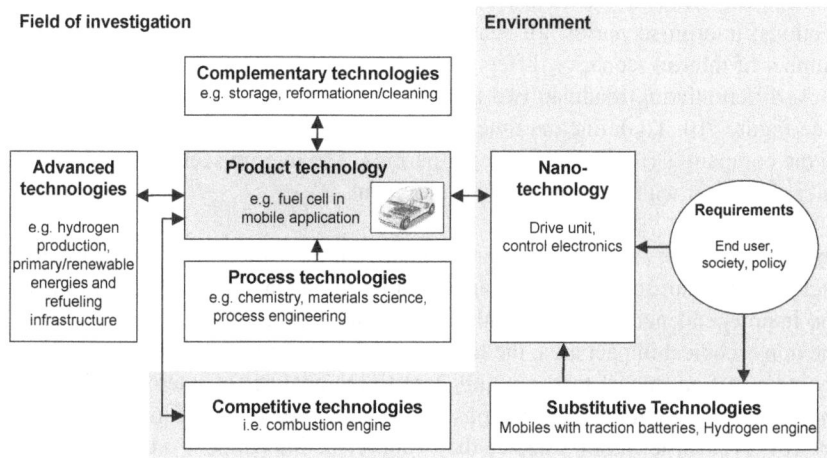

Fig. 6 Technology complex and external impact area on the technological field "fuel cell"

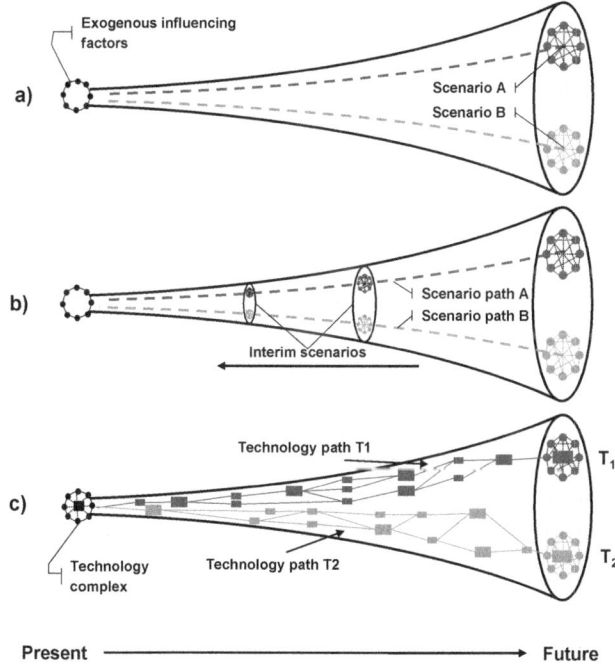

Fig. 7 Scenarios for exogenous influences (a), development of interim scenarios (b) and showing technological paths of development in stages (c)

Generating interim scenarios: Long term scenarios should be divided in time sections; interim scenarios are generated backwards to the present situation. The number of interim scenarios differs according to the given time horizon. However, for a differentiating roadmap two to three interim scenarios should be elaborated (see Figure 7b). To define the time-span of the interim scenarios one should refer to the company's (mid-term) planning periods. The interim scenarios facilitate the later alignment within the strategy development.

Impact analysis: Apart from their influence on the examined technology, some factors in the impact area significantly influence one another. In order to identify the interdependencies existing within the technology complex, as well as within the non-technical impact area, the factors and their relative influences are analyzed by means of an impact matrix, equivalent to the analysis of interdependencies in the technology complex. All relations between the projections of impacting factors are reviewed in terms of whether they support or impediment one another. This analysis does not only facilitate judging which of the factors are the most important "drivers" in the system and thus essentially determine the development of the impact area, but also identifies the factors having a strong impact on the technology while being rather "driven" by the system.

The evaluation of the interdependencies regarding all impacting factors, and the elaboration of the respective projections in addition to the consistency matrix is very time consuming. In order to simplify the work one should not consider all projections in the impact matrix, but only those scenarios that are selected as a basis for the roadmaps.

Stage 3: Developing a roadmap

In the course of this stage, detailed requirements concerning the technology complex and the product technology are derived from the previously established scenarios of the non-technical impact area.

For each scenario corresponding technology roadmaps are developed (Saritas and Aylen, 2010). For this purpose based on the development in the exogenous influencing area, the impact concerning the technology complex is described and specific requirements on the product technology are elaborated. According to this input, the actual state of the product technology is described for the interim point in time. This description also includes the upstream, process and complementary technologies that are required for the entire product technology.[2]

---

[2] Due to the fact that there are alternative developments in the non-technical area, one could also describe alternative projections. In this case a consistency analysis should be carried out regarding the technical impact area for a defined interim-point. Combined with the impact analysis in- and outside the technology complex, the most consistent "technology scenario" serves as the basis for the development of the next interim-scenario.

This procedure has to be repeated for every single interim scenario until the target scenario is reached. This means that for each interim scenario a "new" technology complex has to be described. Technology-specific impacts have to be established and requirements concerning the product technology should be defined.

The development of technology paths starts with the present. The technology pathways are derived from the requirements. There is a sequence of different developments of technologies to fulfill these requirements. Differences in the possible development of the technologies may lead to a ramification of technology paths. In this case the solutions of each branch should be analyzed separately. The time segmentation by interim scenarios determines only cornerstones of the technological developments. If, for instance, the given forecasts concerning the availability of certain technologies are rather uncertain, a time-span should be provided. Thus, the technological developments follow a specific chronological structure differing from that of the interim scenarios.

However, it is important to consider only a few branches. Moreover, the technology pathways should neither include the latest technical developments nor the earliest possible time levels, but it is important to integrate the technology paths into the external impact area. It should be elaborated in detail how the external impact area effects the technology development. Hence, each scenario should be considered separately (Specht, Mieke 2005).

The results obtained in the previous steps are visualized by means of clear graphic illustrations. The main qualitative and quantitative trends of the non-technical impact area which characterize the basic conditions of the scenario are visualized. The depiction of the main performance and diffusion indicators with concrete numbers follows. Finally, the cross-links, i.e. the roadmap itself, are drawn. All indicators are shown in the map in their chronological order. In this final step, the results of all analyses are summarized. The development of the technology at the micro and macro level should clearly be separated.

Figure 8 presents a possible technology roadmap for a fuel cell on the micro level. Each bar corresponds to a development level of the respective technology, drawn in chronological order. The white bars represent the development period of the technology. The grey bars mark the time-span during which the technology is available for utilization. The black bars show the period in which the technology is terminated, e.g. when it becomes obsolete. The arrows stand for structural relations between the technologies, i.e. a technology from which an arrow starts is urgently needed by the technology it points to. In addition, it is possible to visualize outstanding or trend-setting developments (projections) in the non-technical impact area (e.g. introduction of impacting laws).

During the entire process of developing the roadmap, all new findings are constantly compared with the results that have already been produced, in order to attain a consistent picture of the development road. The roadmapping process is not a linear one but rather a permanently iterative process.

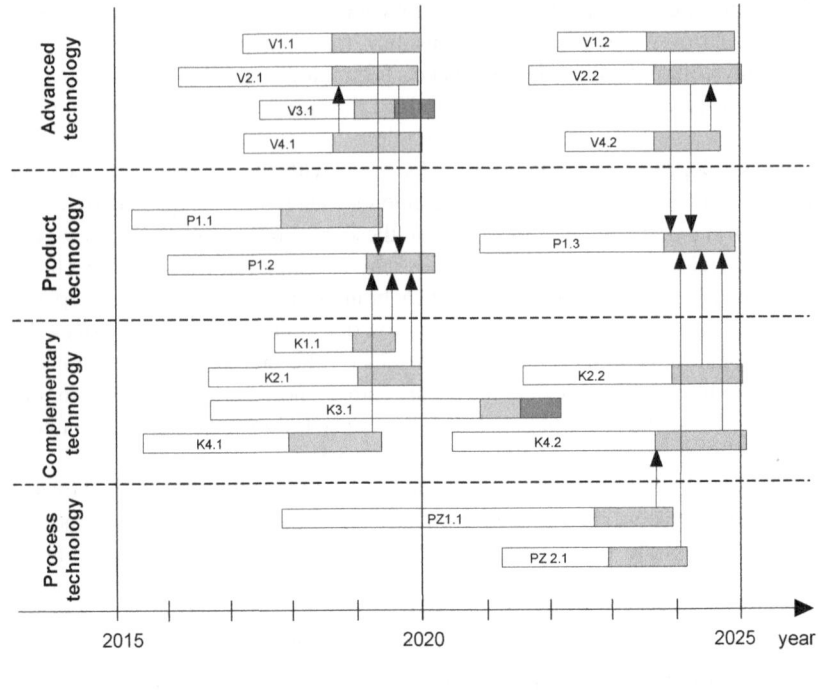

Fig. 8 Technology roadmap for fuel cell application on a micro level

4 Summary

The development of scenario-based technology roadmaps is a very laborious process. Besides, this type of roadmap does not even produce concrete technology and project plans for the company to work with. These remain to be elaborated subsequently to the roadmap. Furthermore, the roadmap has to be reassessed in regular intervals and, if necessary, adapted to the current situation and new insights.

Nevertheless, scenario-based technology roadmaps provide a profound and comprehensive basis for specific planning of technologies and innovations. They represent a method of technological forecasting which enables a company to analyze technologies in a very dynamic environment, for instance, or technologies with high potentials, but not yet well-known. Since the development of scenario-based technology roadmaps is very expensive, SMEs should preferably produce them in form of co-operative projects.

The analysis of interdependencies and dependencies between external developments in the impact area and the progress of a technology also provides a monitoring instrument for strategic technology planning. Roadmaps allow a continuous identification and survey of external influencing factors that determine the development of a given technology (in particular drivers), and thus, if necessary, enable appropriate reactions to actual changes.

References

Behrendt, S., Erdmann, L.: Integriertes Technologie-Roadmapping zur Unterstützung nachhaltigkeitsorientierter Innovationsprozesse. WerkstattBericht, vol. (84). Institut für Zukunftsstudien und Technologiebewertung, Berlin (2006)

Damrongchai, N., Tegart, G.: Future fuel technology scenario and roadmapping for Asia-pacific. In: Second International Seville Seminar on Future-Oriented Technology Analysis: Impact of FTA Approaches on Policy and Decision-Making, Seville (2006)

Fink, A., Siebe, A.: Handbuch Zukunftsmanagement - Werkzeuge der strategischen Planung und Früherkennung. Campus Verlag, Frankfurt (2006)

Garcia, M., Bray, O.: Fundamentals of Technology Roadmapping. Strategic Business Development Department, Sandia National Laboratories, Albuquerque (1997), http://www.sandia.gov/PHMCOE/pdf/Sandia'sFundamentalsof Tech.pdf (access July 12, 2011)

Geschka, H.: Die Szenario-Technik in der strategischen Unternehmensplanung. In: Hahn, D., Taylor, B. (eds.) Strategische Unternehmensplanung - Strategische Unternehmensführung, 8th edn., pp. 518–545. Springer, Würzburg (1999)

Geschka, H.: Technologieszenarien - ein Analyse- und Planungsinstrument des Technologiemanagements. In: Zahn, E. (ed.) Technologiemanagement und Technologie für das Management, pp. 153–171. Schäffer-Poeschel, Stuttgart (1994a)

Geschka, H.: Methoden der Technologiefrühaufklärung und Technologievorhersage. In: Zahn, E. (ed.) Handbuch Technologiemanagement, pp. 623–644. Schäffer-Poeschel, Stuttgart (1994b)

Geschka, H., von Reibnitz, U.: Zukunftsanalysen mit Hilfe von Szenarien - erläutert an einem Fallbeispiel "Freizeit im Jahr 2000". In: Politische Didaktik, vol. (4), pp. 21–101 (1979)

Geschka, H., Schauffele, J., Zimmer, C.: Explorative Technologie-Roadmaps - Eine Methodik zur Erkundung technologischer Entwicklungslinien und Potenziale. In: Möhrle, M.G., Isenmann, R. (eds.) Technologie Roadmapping, 2nd edn., pp. 161–184. Springer, Heidelberg (2002)

Groenveld, P.: Roadmapping Integrates Business and Technology. Research - Technology Management 50(6), 49–58 (1997)

Kostoff, R.N., Schaller, R.R.: Science and technology roadmaps. IEEE Transactions of Engineering Management 38(2), 132–143 (2001)

Oliveira, M.G., Rozenfeld, H.: Integrating technology roadmapping and portfolio management at the front-end of new product development. Technological Forecasting & Social Change 77, 1339–1354 (2010)

Phaal, R., Farrukh, C.J.P., Probert, D.R.: Technology Roadmapping - A planning framework for evolution and revolution. Technology Forecasting & Social Change 71(1-2), 5–26 (2003)

Saritas, O., Aylen, J.: Using scenarios for roadmapping: The case of clean production. Technological Forecasting & Social Change 77, 1061–1075 (2010)

Specht, D., Mieke, C.: Szenariobasiertes Technologie-Roadmapping in Technologiefrühaufklärungsnetzwerken. In: Gausemeier, J. (ed.) Vorausschau und Technologieplanung, Paderborn, pp. 215–242 (2005)

Vinkemeier, R.: Roadmapping als Instrument für strategisches Innovationsmanagement. Technologie & Management 48(3), 18–22 (1999)

Authors

Horst Geschka is a pioneer in the fields of innovation management and creativity enhancement in companies in Germany; with his team he has developed several techniques and instruments used in innovation management including creativity techniques, scenario development, evaluation concepts. Until 2008 Geschka held a chair of entrepreneurship and innovation management at the Technical University of Darmstadt but is still lecturing at other Universities in Germany and Switzerland. He is Management Director of Geschka & Partner Unternehmensberatung and is chairman of the board of trustees of the Fraunhofer INT.

Heiko Hahnenwald is working for Geschka & Partner Unternehmensberatung since 2002. His work focuses on scenario building and technology roadmapping for strategic scenario-based planning with emphasis on innovation and technology management. He also works on projects to design the early phases of the innovation process in companies. He had done a number of scenario studies and technology roadmaps in the areas of mobility and transport, construction and real estate, security and emerging technologies for companies and research institutes. In addition he is further developing the scenario building process and the scenario software INKA.

TRIZ-Based Technology Roadmapping

Martin G. Moehrle

In companies there are many different variants of technology roadmapping in use (see the different chapters in this book). Each variant requires a specific process, in order to forecast future technologies concerning content and time. For this purpose tools from the theory of inventive problem solving are suggested in the following essay, in particular trends of technical systems evolution. They do not replace conventional creative thinking, but supplement it, and lead it in promising directions. Combined with a comprehensive process for the technology roadmapping they unfold their full effect.

1 Technology Roadmapping as an Important Field of an Interdisciplinary Technology and Innovation Management

Technology roadmapping constitutes an important field of an interdisciplinary technology and innovation management. Within such a management a suitable connection needs to be created between potentials, which are offered by the technologies, and needs, which are denoted by markets.

Different functional areas of a company have to be included, starting with research and development (R&D) and marketing, but also production, purchase and finance. Beyond that, it is more and more necessary to build up networks between companies to connect special know-how profitably (see Figure 1; see Rothwell, 2002 about the challenges of innovation management of the fifth generation).

A connecting link between all the outlined instances is built by technology roadmapping, as it is seen in recent time. It can be used in different situations, e.g.

- to coordinate the tactical planning in marketing and R&D of a company,
- to discuss co-operation with another company,
- to develop long-term relations with customers.

Martin G. Moehrle
IPMI - Institut für Projektmanagement und Innovation, Universität Bremen,
Wilhelm-Herbst-Str. 12, 28359 Bremen,
Germany
e-mail: martin.moehrle@innovation.uni-bremen.de

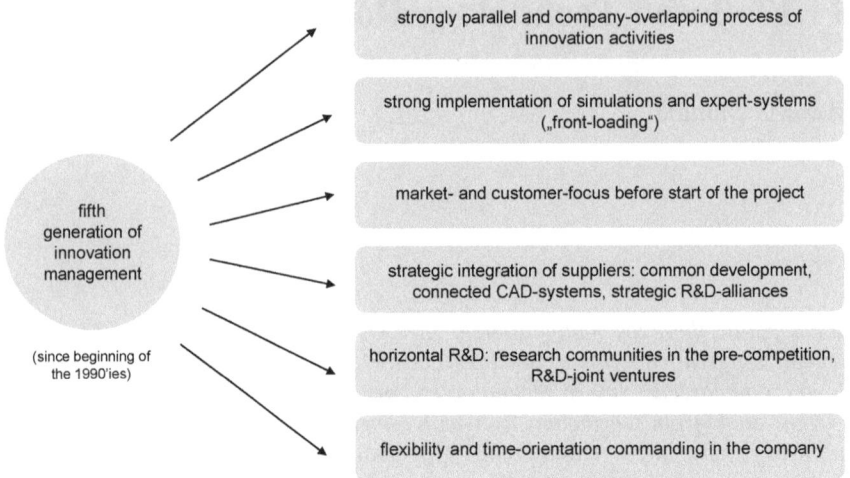

Fig. 1 Challenge to an innovation management of the fifth dimension. Source: content from Rothwell (2002), own diagram.

In any case the forecast of technical systems is a critical point. Usual approaches for this task are associative and systematic creativity techniques, frequently based on analogical formation to well-known systems, or the exhaustion of the employees' know-how of a company as well as of external experts. In the following a basic approach of technology roadmapping is introduced. Additionally to the mentioned approaches this one contains a process developed on the theory of inventive problem solving (commonly shortened as the Russian word TRIZ). This contribution contains two aspects:

- Firstly, parts of TRIZ relevant for technology roadmapping are outlined. In doing so some characteristics of TRIZ are mentioned: (i) its comprehensive cover of inventive problems, (ii) the underlying approach as well as (iii) the extensive experience-based knowledge, it offers.
- Secondly, the link between TRIZ and the process of technology roadmapping will be described. Most notably, trends of technical systems evolution help forecasting future technologies. Derived from this they also help to gain product, process and service ideas.

2 Theory of Inventive Problem Solving

Technical applied sciences, e.g. electronics, thermodynamics, process engineering, mechanical engineering, air and space technology, provide theoretical basics, models and methodologies to solve problems in its fields. With his Theory of Inventive Problem Solving the Russian researcher Genrich S. Altshuller (1926 to 1998) tried to generalize over all these applied sciences i.e. to find theoretical basics, models and methodologies, which are used in all technical applied sciences.

2.1 Basics of TRIZ

In common of all technical applied sciences – and therewith as a central term of TRIZ – the invention abounds in terms of a novel solution of a problem. Altshuller examined numerous of such inventions on the basis of patents and similar property rights. He came to two conclusions:

Firstly, all inventions can be characterised by the contradiction, which they help to overcome. Thereby a contradiction on technical level consists of two functions working in opposite directions: One function is required, but in the moment, in which the desire is achieved in conventional way, the other function changes inadmissible. The contradictions can be standardized by a classification of the desired functions, an analogous classification of the unwanted functions as well as the combination of both.

Secondly, despite all varieties of technical inventions substantial similarities can be identified. Altshuller (1998, p. 186-193) established among others eight very abstract formulated laws of technological system evolution. He also formulated forty substantial more concrete inventive principles (Altshuller, 1998, p. 131-149) and furthermore he developed a multiplicity of separate tools (see the overview in Moehrle, 2005, Mann, 2002, and Pannenbäcker, 2001). From the laws of technological system evolution and the inventive principles evolutionary patterns, suitable for technology roadmapping, aroused later (Figure. 2).

In the first instance the Theory of Inventive Problem Solving has describing character, only in a small measure the character is explaining. Nevertheless the cognitions gained can be transcribed in a constructive way, so additionally an action-leading character results.

2.2 Laws of Technological Systems Evolution

Technical systems follow certain lines of evolution. If these lines of evolution are pronounced over all ranges of technical applied sciences and if necessary also refutable in the sense of Popper, one can call them laws of evolution.

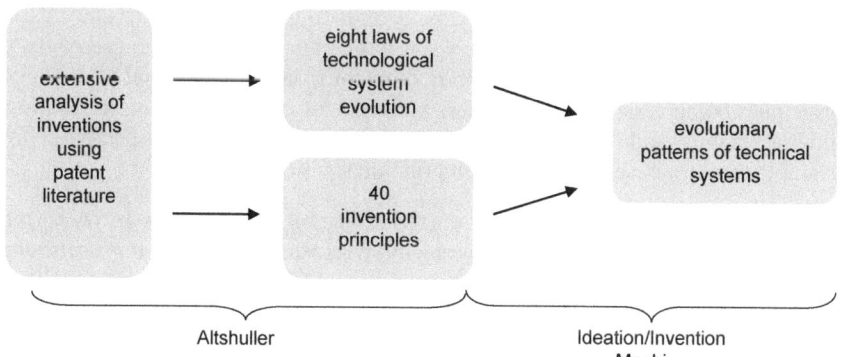

Fig. 2 Coherence of laws of technological systems evolution, the inventive principles and evolutionary patterns of technical systems

Naturally such laws of evolution must be formulated very abstractly. Altshuller (1998, p. 186-193) entitles eight of such laws, by which two are picked out exemplarily:

- Law of uneven development of system parts: "The development of system parts runs unevenly; the more complicated a system, the more unevenly is the development of its parts"(Altshuller, 1998, p.191). The law can be reconstructed by the development of television sets in a descriptive way.

    In the 1980er years a television set consisted of different electronic units, a high voltage unit, a picture tube together with deflection elements and a box. Since then the electronic units have moved in the direction of higher integration and fulfilment of new functions, while the other units remained more or less unchanged. Recently the display unit will change the picture tube will be replaced by flat screens.
- Law of transition to a super-system: "After exhaustion of its development possibilities a system becomes part of a super-system: The further development occurs on the level of the super-system " (Altshuller, 1998, p.191). For this law the cellular phone is a current example. After years of continuous improvements now the time seems to have come, to enhance cellular phones on the level of super-systems, e.g. PDA's (personnel digitally assistants) or multi-communication devices. The laws of technological system evolution, like many other tools of TRIZ, rather inspire to ask, than to give ready-made answers.

2.3 Invention Principles

The 40 invention principles are essentially more concrete and closer to application than the laws of technological systems evolution. They are often arranged in up to five sub-methods (see Altshuller, 1998, p. 131-145). These are heuristics, which intuitively have been used again and again by inventors, in order to solve technical problems. In that a single invention principle does not constitute an outstanding surprise and less than ever no "magic charm" (which sometimes is suggested by advertisement of software producers). Nevertheless with the intuitive use of the invention principles - and mainly with the unconscious use – only few methods turn out, which are used in a company several times (see Moehrle and Lessing 2004 with a relevant comparison between three companies). So the use of the 40 invention principles insists in their comprehensiveness: The use of invention principles being less common so far, activates to break out from conventional thinking habits and assists thinking in different invention perspectives. Two examples may represent the invention principles:

- Principle of conversion from harmful into useful, sub-principle b: "A harmful factor has to be eliminated by overlaying with other harmful factors (Altshuller, 1998, p.139). The active noise damping is an example, where this method is applied. Disturbing noise is antagonised in the following way: (i) nearly the same noise is produced again, (ii) this noise, compared to the original, is displaced out of phase (i.e. delayed temporally) and (iii) the original noise is overlaid by the new one. The result is a clear decrease of the original noise.

- Principle of feedback, sub-principle a: "A feedback has to be introduced" (Altshuller, 1998, p. 139). One can find numerous examples of this principle, for example the automatic brake system in the automotive engineering, by which the traction of a car is optimised by a feedback of each road condition.

Linde and Hill (1993) extended the list of 40 invention principles, suggested by Altshuller, by six further invention principles; whereas Zobel (2001, p. 167) made a containment of 15 universal principles and subordinated the remaining invention principles to these universal ones. Newer descriptions can be found in Herb, Herb and Kohnhauser (2000) as well as in the module "Principles" of Invention machine (2000).

2.4 Evolutionary Patterns of Technical Systems

Evolutionary patterns of technical systems seize beyond the heuristics of invention principles, however they by far do not have the requirement on irrefutability like the laws of technological system evolution have. They are used particularly in software products of Invention machine (2000) as well as Ideation (2000) and were discovered by expansion of invention principles and by concretising some laws of technological system of evolution. Two examples of such evolutionary patterns are mentioned in the following:

- Mono-systems have been developed over the years to bi- and poli-systems (Figure 3). For example by-and-by a stitching head of a sewing machine was upgraded by a folding arm and later by a cutter up to a universal head.

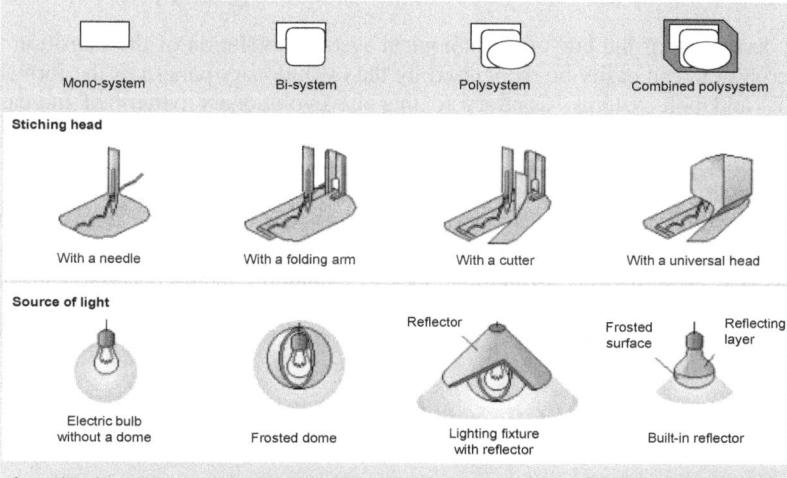

Fig. 3 "Creation of bi- and poli-systems from mono-systems" as an example for an evolutionary pattern of technical systems. Source: Invention Machine (2002), module "Prediction".

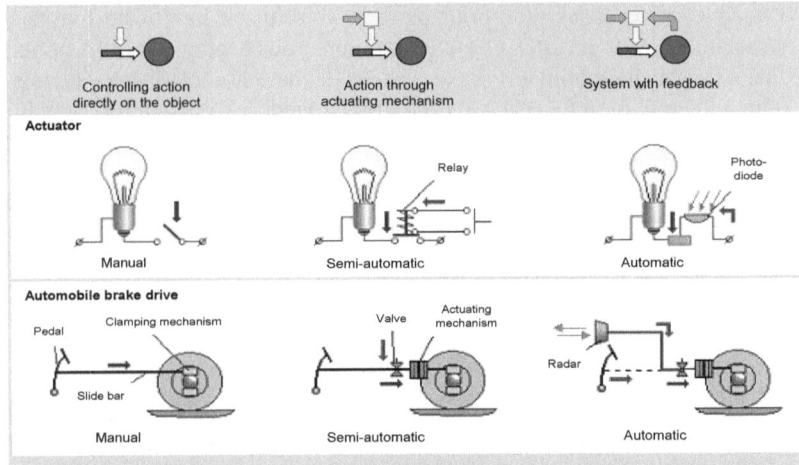

Fig. 4 "Increasing regulation extent" as example for an evolutionary pattern of technical systems. Source: Invention Machine (2001), module "Prediction"

- The regulation extent in technical systems rises (Figure 4). Think of the braking of a car. Today the driver has to brake manually, but if one have an appropriate control variable (e.g. distance measuring by radar beams) a control will be possible by a correcting variable (e.g. brake pressure).

The derivation of the law of technological systems evolution of the transition to a super-system can easily be recognised by the evolutionary pattern of the formation of bi- and poli-systems. Contrary to this the evolutionary pattern of increasing regulation extent directly ties in with the invention principle of feedback. Of course evolutionary patterns of technical systems are not irreversible, leaps in the opposite direction, for example from a poli- to a mono-system, are absolutely possible.

The evolutionary patterns, contained in software products, seem elaborate concerning structure and composition, so in many companies they are used frequently. Nevertheless just here the lack of explaining theory becomes especially apparent. In this respect one cannot speak of a final assembly at all, and the technology roadmappers are held to design and insert further patterns if necessary.

3 Approach to TRIZ Based Technology Roadmapping

The just outlined TRIZ and above all the evolutionary patterns of technical systems can help technology roadmapping in a substantial way. In the following a process is suggested, which contains five steps and it can be processed by department-internal as well as department-external or company-overlapped groups (Figure 5):

- Step 1: Definition of the investigation field
- Step 2: Functional abstraction of the considered system

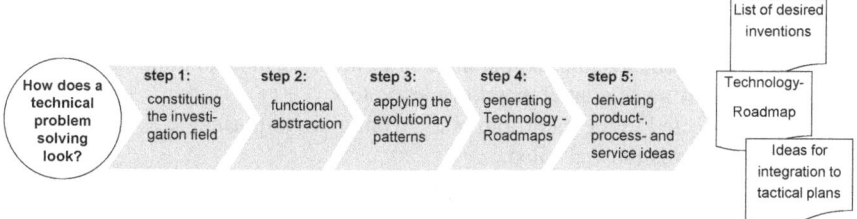

Fig. 5 Approach to the TRIZ-based technology roadmapping

- Step 3: Projection by application of evolutionary patterns of technical systems as well as evaluation of relevant technologies
- Step 4: Creation of technology roadmaps
- Step 5: Derivation of product-, process- and service-ideas by technology roadmaps

A continuous example from the sanitary sector may illustrate the approach.

Both, step 2 and step 5 in parts form an application of morphologic thinking according to Zwicky (1989). This will be described explicit.

3.1 Step 1: Definition of the Investigation Field

Firstly, the investigation field for technology roadmapping has to be specified. Therefore three starting points are differentiated:

- Is there a certain technology, which one can focus on, independent of the providers in this field? If so, technology roadmaps giving an orientation about possible chances and risks to the management of a company will result (see the examples to the photovoltaic and gas cell in the contribution of Geschka and Hahnenwald in this book).
- Shall be focused on several technologies, which occur together in a certain application system, also independent of the providers in this field? If so, technology roadmaps giving orientation over possible chances and risks to the management of a company will result in the same way.
- Or should the power spectrum of a company or a division be considered? In this case product-roadmaps are necessary beneath technology roadmaps (see the integrated example in the contribution of Specht and Behrens, 2002).

In each case the actual state of the selected system should be documented firstly.

In the example of the sanitary sector a conventional shower cubicle, how it can be found in nearly each household, is selected as investigation field. The object of technology roadmapping is to create ideas for a certain company, which widely go beyond the pure improvement of today's products. These ideas are derived from a forecast for the entire application system, which is relevant to companies of different industrial sectors. The today's condition of a shower cubicle can be characterised roughly with eight points:

- Water running constantly, likewise water temperature, after initial cooling shock,
- Water requirements and temperature calibration by manual controls
- Base consisting of coated metal or plastic tub,
- Lining directed to the bathroom made from glass or plastic,
- Lining towards the wall tiled,
- Bought cleaning supplies for person and items within the room,
- Drying process of persons by towels,
- Manual cleaning of fittings, tub and cubicle necessary.

3.2 Step 2: Functional Abstraction of the Considered System

In step 2 the considered system has to be split up into its current functions and the functions that are desirable in the future. This step is based on morphologic thinking according to Zwicky (1989, p.116): by splitting up an item into stand-alone components creativity can already be stimulated (also see Pahl et al., 2007 for the deployment of morphologic thinking in the design of engineering and Moehrle 2010 for the combination of morphology with TRIZ). While Zwicky, being an astrophysicist, might rather have had technical problems in mind, his ideas were adopted to the marketing especially the conjoint measurement and they were enriched by a customer-oriented perspective (see Gustafsson et al., 2007 regarding the aspect of conjoint measurement). Insofar a functional abstraction from customer view should be aspired, in order to achieve marketable solutions. The suitable question is: "Which functions do customers expect from a system respectively a technology?" and not "Which (technical) function does the system respectively the technology cover?" For the technology roadmapper this also implies an imagination about the question which customer groups should be addressed today and in future.

Again applied to the example of the sanitary sector this quickly leads to the main function of certain "wellness"-feelings wished by the customers: At cold outdoor temperatures they want to feel well-warmly-cleanly after taking a shower, at warm outside temperatures they want to feel freshly-cleanly. The main function then can be subdivided into further functions (Figure 6).

3.3 Step 3: Projection by Use of Evolutionary Patterns of Technical Systems

For the individual functions extracted from the overall system in step 2 now projections have to be generated. This also is usual to the morphologic thinking and in the simplest case it is done by brainstorming of experts (see the overviews of Geschka and Dahlem, 1996 as well as Hauschildt and Salomo, 2010; furthermore Isaksen et al., 2010 to a recommendable form of brainstorming). Additionally to the brainstorming it is proposed in this essay to use evolutionary patterns of technical systems already outlined in this essay. The intention is to focus the creativity of the technology roadmapper precisely at important technology-overlapping trends.

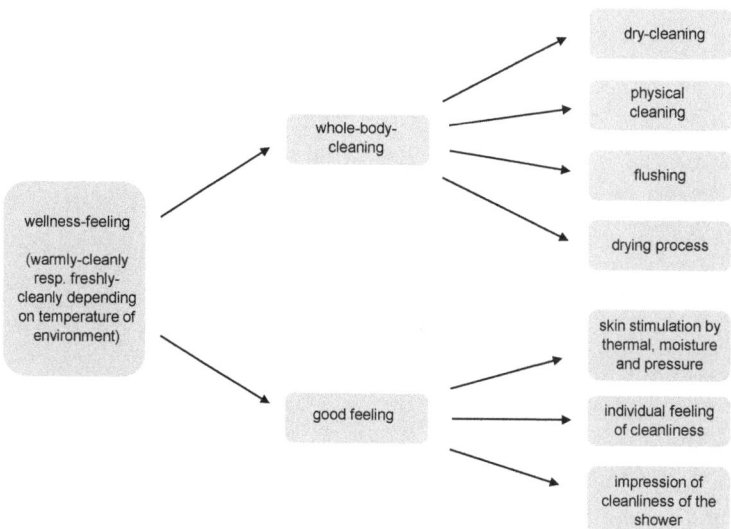

Fig. 6 Functional abstraction (exemplified with the example of a conventional shower)

In the operational conversion step 3 of the suggested process covers a mini loop, which can be passed by the technology roadmapper in two variants.

- Variant a: The technology roadmappers select a first function of the application system and apply to it different evolutionary patterns of technical systems. Then they select the next function and repeat this process, until all functions are processed.
- Variant b: The technology roadmappers select a first evolutionary pattern and apply it to all functions of the application system. Then they select the next evolutionary pattern and repeat this process, until all evolutionary patterns are processed.

Hybrid forms from both variants are also conceivable, of course. In each case it is advisable to consider two aspects: on the one hand the evolutionary patterns should be deliberated completely. If using the evolutionary pattern of increasing regulation extent in technical systems for example, the technology roadmappers shall not only mention possible correcting variables, but also underlying controlled variables (see example below). On the other hand the technology roadmappers should describe and evaluate ideas, arising by appliance of evolutionary patterns, concerning three main and two deepening aspects (Figure 7).

As a result of step 3 a list is developed. This list contains customer weighted ideas, their technical problems to be solved, the time horizon, up to which the appropriate technologies will be available, and finally the competence of the own company.

A small section of the shower example should help to illustrate step 3. Further cut-outs can easily be considered on the basis of laws of technological system evolution (Figure 3 and 4), already mentioned. The function "skin excitation by

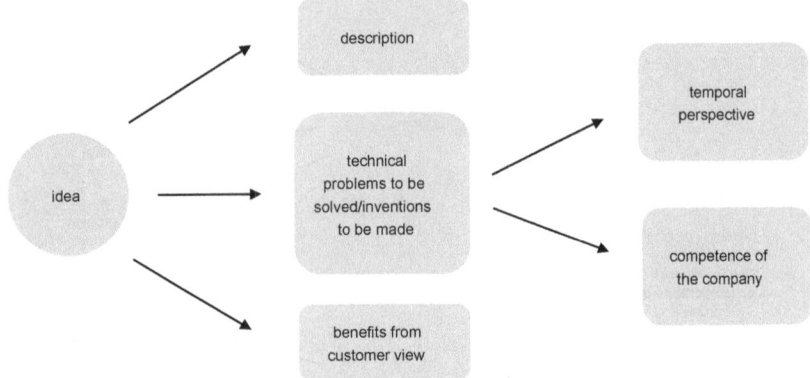

Fig. 7 Criteria for evaluation of the ideas in Step 3 of technology roadmapping

warmth" is projected into the future on the basis of the evolutionary pattern of increasing regulation extent in technical systems (Figure 4). In a testing phase concrete ideas appeared, for example: depending upon outside temperature the temperature transmitter puts up or off the water temperature by two degrees, whereby the scale on the temperature transmitter does not contain degree numbers, but symbols. In addition, more unusual ideas are possible: depending on the cleanliness of the drain water more or less cleaning supplies are given to the shower water. Here from technical view an interesting question is, how to measure the cleanliness of the drain water physically or chemically - and in relation to this how to measure the cleanliness of the person taking a shower. This is a central point in the evaluation, as technical problem to be solved or as an invention to be made (Figure 7). For this it now applies to measure both, a realistic temporal perspective and the company-own competence. Likewise it is to be analyzed to what extent a benefit is recognized by the customer by solving this problem.

3.4 Step 4: Formation of Technology Roadmaps

Step 4 results on the list developed in step 3. Now the technology roadmappers should cluster the technical problems to be solved respectively the inventions to be made to technology fields (in the following called technologies). For these technologies two kinds of information need to be requested:

- On the one hand realization times have to be indicated. For this it can be fallen back to approximations of the technical problems to be solved. Frequently economic aspects get involved here. Thus it is not sufficient that a technology is present in principle (think on space flights). It also has to be that kind of beneficial to be used in considered application system yet (for vacation trips etc.). The indication of the realization times requires large investigations, by asking technology suppliers, research institutes and other suppliers of know-how.

TRIZ-Based Technology Roadmapping

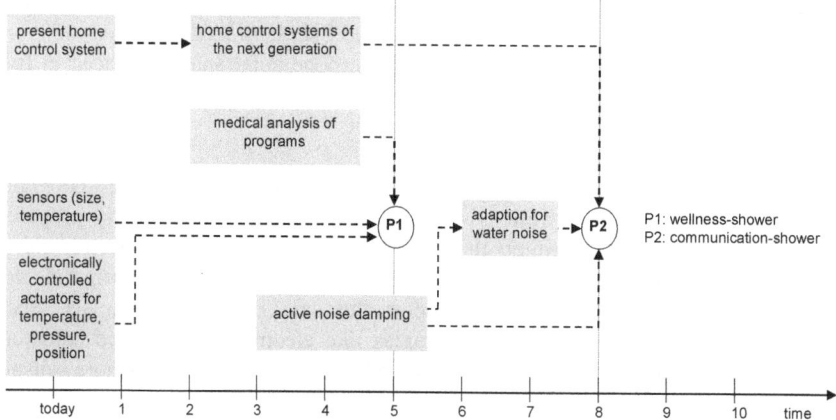

Fig. 8 Part of a technology roadmap

- On the other hand functional dependences between the different technologies have to be specified. Depending on investigation field a relative small number of such functional dependences can occur here.

With the mentioned data the technology roadmappers can produce a technology roadmap now. In doing so further technologies still can be supplemented. Furthermore, from such a technology roadmap it quickly becomes evident, in which aspects the technology roadmappers have to intervene actively by activating R&D projects, and in which aspects they depend on third. Concerning the showering example the technology of active noise control (see the section about the invention principles) was recognised as substantial. This technology will be applicable in some years, though still a technology for water areas has to be developed (Figure 8). Here the responsible persons of a company can intervene actively. Whereas house control systems, whose next generations are already foreseeable, seem little promise to adopt an active role. Here co-operation with leading manufacturers seems to be more promising, the integration of the sanitary area can be designed together.

3.5 Step 5: Derivation of Product -, Process- and Service Ideas from the Technology Roadmaps

Finally, different ideas for new products, processes and services can be derived from the technology roadmap developed in step 4. For this it is a good solution to let fall a perpendicular in the graphic at different times to determine which technologies will be available up to then. Thereafter ideas can be produced by creative combination. This step also ties in with morphologic thinking. With each perpendicular the technologies being presumably available up to then are considered as idea dispensers. Then the strategic R&D as well as marketing planning of a company can be supplied with these ideas (see Burgelman, Christensen and Wheelwright, 2004; Ahmed and Shepherd, 2010).

In the example of the shower cubicle the perpendicular shall be placed at five and eight years, counted from the starting point of operation (midyear 2000) (again see Figure 8). Two product ideas step out:

- In approximately five years the "wellness-shower" will become more and more common. This "wellness-shower" offers different medically prepared showering programs, which the user selects depending upon desire, e.g. a refreshment program, a relaxation program or an energising programs for special parts of the body to support the care of physiotherapists. Such "wellness-shower" requires technologies like electronically steered actuators for water temperature, water pressure and applying place, furthermore sensors for simple parameters (e.g. outside temperature, size and shape of the showering person). Finally, a control unit operating commensurate to medically secured conclusions is needed.
- In approximately eight years communication media also becomes part of the shower, so a chat between the showering person and an outside standing person (equal, if from the same room, from an adjacent room or from the telephone) becomes possible. The water noises such as a pattering or bubbling are faded out to a large extent by means of "active noise control". A connection to the house control system controlling different equipments would also be advantageous.

Of course, ideas for processes and services aid the product mentioned ideas. Thus interesting perspectives for medical supply arise from the first idea. Interesting prospects concerning conversation result by the second idea and one can find new ways of maintenance and renewal of the new showers respectively re-fitting of conventional showers by both ideas.

4 Summary

Technology roadmapping in the outlined form is an effective tool for the conversion of a functional-overlapping and interdisciplinary innovation management. It generally helps the involved persons by consensus identifying and common adjustment in a substantial way. Technology roadmapping can be used in different ways and at different effort. The spectrum ranges from one person a day, i.e. if a director of a R&D unit applies the technology roadmapping to the domain of his unit, up to several person months, if several companies of one industry want to agree on a common technology roadmap for technology-political reasons.

In form suggested here technology roadmapping is based on the Theory of Inventive Problem Solving. From this in first instance the evolutionary patterns of technical systems get applied. Whether working with or without evolutionary patterns: the outlined process in this essay arranges the subtasks of technology roadmapping into five steps, which ensure a systematic completing and comprehensible results.

References

Ahmed, P.K., Shepherd, C.D.: Innovation management. In: Harlow, et al. (eds.) Context, Strategies, Systems and Processes. Prentice Hall (2010)

Altschuller, G.S.: Erfinden - Wege zur Lösung technischer Probleme, 2nd edn. Planung und Innovation, Cottbus (1998)

Burgelman, R.A., Christensen, C.M., Wheelwright, S.M.: Strategic management of technology and innovation, 4th edn. McGraw-Hill, Boston (2004)

Geschka, H., Dahlem, S.: Kreativitätstechniken und Unternehmenserfolg. Technologie & Management 46(3), 106–110 (1996)

Gustafsson, A., Herrmann, A., Huber, F.: Conjoint measurement: methods and applications. Springer, Heidelberg (2007)

Hauschildt, J., Salomo, S.: Innovationsmanagement. Vahlen, Munich (2010)

Herb, R., Herb, T., Kohnhauser, V.: TRIZ - der systematische Weg zur Innovation. Moderne Industrie, Landsberg am Lech (2000)

Ideation International (ed.) Innovation WorkBench 2000 (software product), Detroit, Michigan (2000)

Invention Machine (ed.) TechOptimizer Professional Edition 4.0 (software product), Boston, Massachussetts (2002)

Isaksen, S.G., Treffinger, D.J.: Creative approaches to problem solving: a framework for innovation and change. Sage, Thousand Oaks (2010)

Linde, H., Hill, B.: Erfolgreich erfinden. Widerspruchsorientierte Innovationsstrategie. Hoppenstedt, Darmstadt (1993)

Mann, D.: Hands-on systematic innovation. CREAX, Ieper/Belgium (2002)

Moehrle, M.G., Lessing, H.: Profiling technological competencies of companies: a case study based on the Theory of Inventive Problem Solving. Creativity and Innovation Management 13(4), 231–239 (2004)

Moehrle, M.G.: What is TRIZ? From conceptual basics to a framework for research. Creativity and Innovation Management 14(1), 3–13 (2005)

Moehrle, M.G.: MorphoTRIZ - Combining morphological and contradiction-oriented problem solving to solve vague technical problems. Creativity and Innovation Management 19(4), 373–384 (2010)

Pahl, G., Beitz, W., Schulz, H.-J., Jarecki, U., Wallace, K., Blessing, L.T.M.: Engineering design: a systematic approach, 3rd edn. Springer, London (2007)

Pannenbäcker, T.: Methodisches Erfinden in Unternehmen. Bedarf, Konzept, Perspektiven für TRIZ-basierte Erfolge. Brandenburgische Technische Universität Cottbus: Dissertation (2001)

Rothwell, R.: Towards the fifth-generation innovation process. In: Henry, J., Mayle, D. (eds.) Managing Innovation and Change. Sage, London (2002)

Specht, G., Behrens, S.: Strategische Planung mit Roadmaps. Möglichkeiten für das Innovationsmanagement und die Personalbedarfsplanung. In: Moehrle, M.G., Isenmann, R. (eds.) Technologie-Roadmapping. Zukunftsstrategien für Technologieunternehmen, pp. 85–104. Springer, New York (2002)

Zobel, D.: Systematisches Erfinden. Methoden und Beispiele für den Praktiker. Expert, Renningen (2001)

Zwicky, F.: Entdecken, erfinden, forschen im morphologischen Weltbild, 2nd edn. Baeschlin, Glarus (1989)

Author

Martin G. Moehrle is director of the institute for project management and innovation (IPMI) at the University of Bremen, Germany, since 2001. From 1996 to 2001 he was leading the chair for planning and innovation management at the Technical University of Cottbus, Germany. The major research interests of IPMI are technology forecasting, evaluation of innovations, patent strategies, and TRIZ. Prof. Moehrle has published several books and articles in the field of MOT.

Development of Technology Foresight: Integration of Technology Roadmapping and the Delphi Method

Daisuke Kanama

This study examines the integration of the Delphi method with technology roadmapping as a new technology foresight process. The Delphi method and technology roadmapping have developed in different ways, and these two foresight methods are now attracting attention from both national governments and private companies. However, each method also has limitations when attempting to deal with accelerating technological complexity and sophistication and latent markets. This study reviews the merits and demerits of the Delphi method and technology roadmapping, and proposes a new method of technology foresight, which takes advantage of the strengths of both methods.

1 Introduction

In recent years, the importance of innovation in socio-economic development has become widely recognized. Today's innovation research devotes attention not only to R&D in universities and private companies, but also to the importance of constructing a national innovation system that effectively generates innovation in comprehensive systems incorporating market and social needs, institutions, and the regulatory system (Goto and Odagiri, 2003; Goto and Kodama, 2006). At the same time, various issues have arisen, such as latent markets, social needs, and the emergence of more sophisticated and complex technology.

Recent years have seen that the distinction between science and technology become less pronounced. According to Stokes (1997), research can be categorized into three types (see Figure 1). Bohr-type indicates theory oriented pure basic research because Niels Bohr, born in 1885 and won the Nobel Prize in physics 1922, contributed to the theoretical investigation in the development of quantum physics. Application oriented research is called Edison-type since Thomas Edison was a great inventor and businessman who developed many devices that influenced life around the world. Louis Pasteur was a chemist and microbiologist to be

---

Daisuke Kanama
Department of Business and Information Systems, Hokkaido Information University,
Nishinopporo 59-2, Ebetsu, Hokkaido,
069-8585, Japan
e-mail: dkanama@do-johodai.ac.jp

remembered for his remarkable works in the causes and preventions of diseases. Pasteur-type is considered as use-inspired basic research, introducing the features from both Bohr-type and Edison-type. Pasteur-type research has gained in importance recently because innovation has become increasingly science-based in recent years (Kondo, 2009). In other words, the most advanced technology cannot be realized without science-based knowledge, and consequently, this fact affects research in universities and public research institutes. On the other hand, even basic research in universities cannot be performed without considering social needs.

The broad aim of technology foresight is to identify the emerging generic technologies, which are quite likely to yield the greatest economic and social benefits. During the 1990s, technology foresight became much more widespread (UNIDO, 2005). Since 2000, most advanced technologies have become progressively more sophisticated and complex, and as such, the risks inherent in those technologies have also increased. Many firms that had previously conducted R&D internally are now outsourcing all of their R&D tasks and the associated risks, except for their core technologies (Chesbrough, 2003). Complex technologies are now narrowing the distance between science and technology. In universities and national research institutes, for example, public R&D strategies, which were previously considered pre-competitive research, are now influenced by the needs of large firms and market strategies. In these circumstances, technology forecasting has become more difficult. A wide range of foresight methods are available; some are specifically designed for future work, while others are developed in management and planning. It is important that the methods chosen from the available range be suitable for their intended purpose. Exploring possible, probable, and preferable futures relies on assumptions about the future and how we relate to it, which in turn will influence the choice of methods (UNIDO, 2005). However, technology roadmapping, scenario planning, and the Delphi method are the most common and available processes. As methods, roadmapping and scenario planning are more qualitative than quantitative, while the Delphi method uses statistical techniques to quantify forecasts.

The development of foresight has occurred as a response to changes in the world economy. Some of the main drivers of change in the global economy over the coming decades (Martin, 2001) are expected to be:

- Increasing competition
- Increasing constraints on public expenditures
- Increasing complexity
- Increasing importance of scientific and technological competencies.

These factors also underlie the upsurge of interest in technology foresight, giving rise to its emergence as a global concept and policy tool (UNIDO, 2005).

Technology roadmapping has been performed using the Delphi survey, as exemplified in the 'EU Nanoroadmap' and similar projects. Methods relating the Delphi method and scenario writing have also been proposed (Kameoka et al., 2004; Banuls and Salmeron, 2007). However, these approaches tend to be shallow and limited to specific fields, and numerous issues arise when attempting to

Development of Technology Foresight: Integration of Technology Roadmapping

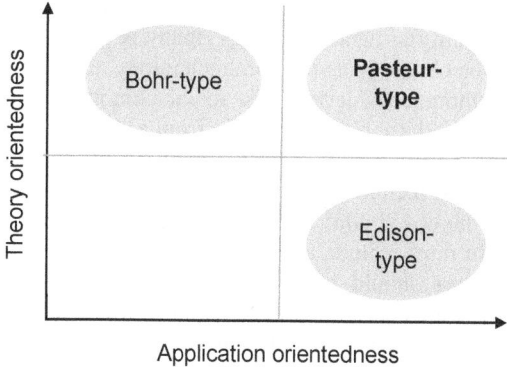

Fig. 1 Three types of categorization of R&D (Based on Stokes, 1997)

integrate multiple technology foresight methods (Caldwell et al., 2005; Kanama, 2007). This chapter examines the potential for integrating the Delphi method with technology roadmapping and enhancing its process to support decision-making, both in public policy and in the management of private companies.

This chapter consists of these sections; Section 2 briefly shows what is needed for technology foresight and how technology foresight should be. Section 3 reviews how the Delphi method has been developed in Japan. Section 4 reviews methodology and function of technology roadmapping, and discusses limitations of technology roadmapping. Then, merits and demerits of Delphi method and technology roadmapping are discussed in Section 5. This section also explores the possibilities of integration of these two methods. Section 6 presents brief overview of the EU Nanoroadmap project as a case study on how technology roadmapping and the Delphi method could be integrated. Finally, integrated foresight guideline is showed in Section 7.

2 Essential Conditions for Technology Foresight

The outcomes of scientific and technical research have implications for society and exert a strong mutual influence on each other. On the other hand, questions have arisen as to the achievements of the science and technology. As a result, technology foresight is no longer simply a technique for forecasting the future of technology, but has also become a tool for creating a vision of the social future, including the ideals and goals of the society, economics, government, etc. (Preez and Pistorious, 1999; Landeta, 2006). From this viewpoint, technology foresight should include the following three elements.

First, technology foresight should include not only technical feasibility, but also a variety of social changes and related elements, for instance, institutional arrangements, global competition, environmental problems, culture, educational systems, etc. (Karube, 2001). To give one example, nanotechnology is a typical field of research involving the most advanced technologies and has attracted great expectations and attention as a next-generation technology. However, owing to the opacity and hyper-reality of nanotechnology, many people have misgivings about products using nanotechnology as the core technology and will not readily accept such products. Without considering this social condition, it would be almost meaningless to forecast the achievement time from a strictly technical viewpoint.

Second, in view of the closer relationship between science and technology and society in recent years, technology foresight should encompass a wider range of stakeholders and thereby build a consensus among them. In other words, technology foresight must include a vision of the future society. This implies that the many stakeholders should have an interest in realizing the aimed future society. Whether or not a consensus of stakeholders exists will make an especially important difference in the implementation phase. In particular, the investment requirements for science and technology have risen dramatically in recent years, requiring the participation of stakeholders. Therefore, the necessity of this kind of investment must be clearly understood by stakeholders.

Third, various forms of prior social consensus for technology foresight are required in light of the increased technological uncertainties associated with progress in science and technology, and particularly the increasingly complexity of technology. It is also desirable to maintain a high degree of flexibility after the initial technology forecast. The construction of a technology foresight tool, which enables repeated upgrades of technology forecasts, is needed, given global competition and the extremely high pace of change in international society.

3 Development of the Delphi Method in Japan

The Delphi method is a methodology of repeatedly conducting the same questionnaire survey with a large number of samples, and having the respondents' opinions converge. After the second questionnaire, the respondents are given feedback on the results of the previous survey. One of the main characteristics that differ from usual questionnaires is that the respondents reevaluate their own answers by looking at the overall trend of the opinions (NISTEP 2005a). The survey process chiefly starts from making the Delphi topics, which are matters concerning the science and technology that should be achieved in the future, and questionnaire items like the time of technological realization, the importance, etc. (Eto, 2003; Kuwahara, 2001).

Another characteristic of the Delphi method is the anonymity of the respondents. Information on the details of respondents and their individual answers is not given to the respondents, though they learn the overall trend of the answers from the second time onward. Therefore, an intentional bias resulting from recognizing specific respondents can be excluded (Wounderberg, 1991). The Delphi method usually forecasts technology trends in 20–30 years' time. Normally, the only source that can be relied on for making a forecast for such a long-term span is said to be the opinions of specialists in each field (Wounderberg, 1991).

The Delphi method is used more frequently than before and the methodology has been repeatedly improved in the past ten years or so (UNIDO, 2005). It is no exaggeration to say that Japan has the longest experience in the Delphi method.

The technology foresight survey in Japan originated in the investigation conducted for the first time by the Science and Technology Agency in 1971 (Kuwahara, 1999, 2001). It has been conducted periodically every five years or so since then. The survey methodology follows the Delphi method, which a USA think tank, RAND, developed in the 1950s. However, the Delphi method that had been used in the USA at that time was relatively small in scale, with about a dozen specialists in specific technological areas participating. When adopting the Delphi method, the Science and Technology Agency introduced the methodology of having several thousand specialists who basically cover all technology fields participate in the investigation to reach a consensus (Kuwahara, 1999). From the 1970s to the 1980s, technology foresight was chiefly used in Japan to examine the common goal to be achieved from a long-term viewpoint, with the involvement of the industrial world.

The Delphi survey in Japan has the feature of having been conducted periodically for 30 years at intervals of about five years, maintaining a basic survey design, which has not been seen anywhere else in the world. In the 1990s, major countries in Europe such as Germany, Britain and Finland started technology foresight (NISTEP, 2001, 2005b). For instance, Germany adopted Japan's methodology and Delphi topics for technology foresight immediately after the union of East and West Germany, and translated the questionnaire of the 5th technology foresight survey conducted in Japan into German (NISTEP, 1994).

On the other hand, in the 1990s, the meaning of technology foresight gradually changed in Japan too. The reasons were as follows:

- The importance of science and technology increased in line with economic globalization and the intensification of competition
- As Japan's position in the world changed from catch-up to top-runner, there were calls for innovations based on science and technology
- Along with the slowdown of economic growth, increased expenditures including medical care expenses pressed the government budget, and stricter evaluation of the validity of science and technology investments came to be demanded.

In addition, given the situation described above, a change took place in the system for planning science and technology policy. The Science and Technology Basic Law was established in 1995, providing the framework of Japan's science and technology policy. The Science and Technology Basic Plan, which is a five-year plan, was formulated, starting from the year after the enforcement of the law. The establishment of this basic plan had a strong influence on technology foresight. It is necessary for policymakers to know the specialists' consensus beforehand, because the system of policymaking changed from a bottom-up process to a top-down process. In the examination of the third Science and Technology Basic Plan in particular, a more evidence-based policymaking system was requested, resulting in an increase of the importance of technology foresight. The National Institute of Science and Technology Policy conducted the 8th technology foresight survey project for two years from 2003 (NISTEP, 2005b), and the results of the project were periodically offered to the Council on Science and Technology Policy, which is the organization creating the Science and Technology Basic Plan.

Moreover, in 2006, the 'Innovation 25' project was mounted in order to show the desired vision of Japanese society in 2025 and the route leading to that vision. The 8th technology foresight results contributed useful information to this project (NISTEP, 2007). Consequently, this experience clarified the need for mission-oriented technology foresight that deals with not only technological issues but also social issues. In the 8th technology foresight survey project, the question about the time for realization was divided into the "Technological realization time" and "Social application time", to emphasize the social aspects.

4 Technology Roadmapping

The purpose of technology roadmapping is to visualize technological issues to be solved, products, and markets along a time axis (Phaal et al., 2004, 2005). In this section the development, background and diversification is presented.

4.1 Development of Technology Roadmapping

Roadmaps originated as a management tool for R&D strategy in private companies. The first roadmap that was widely known in society was a roadmap of the company's own technology, which Motorola Inc. published at an academic conference in the late-1980s (Willyard and McClees, 1987). However, the roadmap with the greatest impact was created by the semiconductor industry. This was the International Technology Roadmap for Semiconductors (ITRS), which is the most well-known roadmap in the world. This roadmap was created by an international consortium of the USA, EU, Japan, Korea, and Taiwan, and is now updated every six months and completely modified every two years.

The development of roadmaps in the semiconductor industry implies two important aspects of technology foresight. One is that we can apparently confirm the increasing importance of roadmaps as technology foresight and a vision of the future society. The above examples show the evolution of technology roadmaps from a strategy tool for private companies to an R&D management tool at the national level. The second aspect is that increasingly specific knowledge is needed to build a technology roadmap. This originates from progress in technology and diversification of social needs. Technology roadmaps must be considered as a type of technology foresight, which includes various social elements.

4.2 Background of the Expansion of Technology Roadmaps

There seem to be three major reasons why the roadmaps are now needed by both governments and private companies. First, because of the rapid increase in technological complexity and diversification of market needs, it has become necessary to grasp technological trends and market needs strategically based on technology foresight (Yasunaga and Yoon, 2006). In other words, more efficient and strategic R&D management is needed to survive global competition. Moreover, because of the increase in converging technologies and boundary research areas, more selective and concentrated R&D strategies are needed. This increases the importance of technology roadmapping as an R&D management tool (Petrick and Provance, 2005).

Second, due to the increasing complexity of technology in recent years, it is more difficult to conduct all R&D in only one company or one country (Chesbrough, 2003). As a result, strategic selection and concentration of R&D and greater focus on core technologies is necessary, especially in the semiconductor industry and biotechnology industry, where growth of R&D investment is particularly rapid.

Third, recent global competition in creating innovation now requires clear cost-effectiveness in R&D investments. This is not limited to private business. In national government R&D policy, society also demands a clear estimation of the cost–benefit performance against the budgetary investment. This has also increased the importance of technology roadmapping and foresight (NISTEP, 2010a; NISTEP, 2010b).

4.3 Diversification of Technology Roadmapping

Technology roadmapping is based on specific technological knowledge. As technological complexity increases, the search range of technology seeds becomes wider, which means it is necessary to collect not only publicly available information such as published academic papers and patents, but also essentially confidential information such as know-how and knowledge in industry. However, even if it were possible to obtain all the relevant technological information, building a roadmap would still involve a certain amount of uncertainty and risk owing to the increase in unexpected complexities, interdisciplinary nature of technological areas, and growth of competitive technologies. Furthermore, market needs have become not only more complicated, but also more important. Users' preferences in the market cannot be predicted accurately in advance. Another important consideration is the problem of 'self-fulfilling predictions'. In other words, the roadmap may actually affect the forecasted future trends in both technologies and markets.

Based on the factors that affect roadmaps, as described above, the formula, shape, and components of the roadmap will depend heavily on the purposes of roadmapping. National governments, industry consortiums, and private companies can all be builders of roadmaps. However, even in the same area of technology, these respective entities will use different roadmapping elements, including the timeframe of the roadmap, the scale of investment, and the R&D organization. For example, if a government or public community builds a roadmap, it has to consider national competitiveness and the reasonableness of public investment, and ultimately, the roadmap will be long-term rather than short-term. Table 1 shows major technology roadmaps worldwide. As shown here, in addition to the semiconductor industry, a wide variety of entities at various levels are building roadmaps. Technology roadmapping is beneficial in technology foresight, but certain limitations should be recognized. As many researchers have pointed out, technology roadmapping involves

Table 1 Major technology roadmaps in the world

Name	R&D field	Organization	Time
International Technology Roadmap for Semiconductors (ITRS)	Semiconductor	International Association of Europe, Japan, Korea, Taiwan, and the United States	Every 2 years since 1999
Chemical Industry Vision2020 Chemical Industry R&D Roadmap	Chemistry	American Chemical Society	1996
NIH Technology Roadmap	Life science	National Institute of Health	2005
US Biomass Technology Roadmap	Biomass energy	US Government	2003
NanoRoadMap	Nanotechnology	European Comission	2005
Technology Roadmap	24 R&D fields	Japanese Government	2005

the following two dilemmas: As the purpose of the roadmap becomes more general, the roadmap itself becomes less strategic. Conversely, as strategic characteristics become more prominent, the roadmap will have a narrower technological search range and may not be published. Many published roadmaps are based on the 'highest common factor' to avoid conflicts of interest.

Because most technology roadmaps have a timeframe extending from the present to 10–15 years in the future, it is easier to consider roadmapping by linear thinking. This means that securing 'after-the-fact flexibility' in advance is a necessary process. Establishing a 'self-organizing' technology roadmap may be a future challenge for the technology foresight research community (Martin, 2004). Consideration of the fact that recent R&D uses a more non-linear process is also necessary. These two dilemmas are discussed in detail in the following.

5 Interdependence between Technology Roadmapping and the Delphi Method

The transition in the Delphi survey in Japan and technology roadmapping has been discussed in Sections 3 and 4, respectively. In the following the interdependence between technology roadmapping and the Delphi method will be discussed.

5.1 Merits and Demerits of Technology Roadmapping and the Delphi Method

One of the largest merits of technology roadmapping when compared with the Delphi method is that roadmapping includes R&D targets, an image of the society, a vision of the future, and concepts. The creation of a roadmap begins from the establishment of the concept. The issues are gradually broken down and linked to the element technology topics necessary to achieve the projected timing of realization of a technology. As a result of this process, a consensus among the

Development of Technology Foresight: Integration of Technology Roadmapping

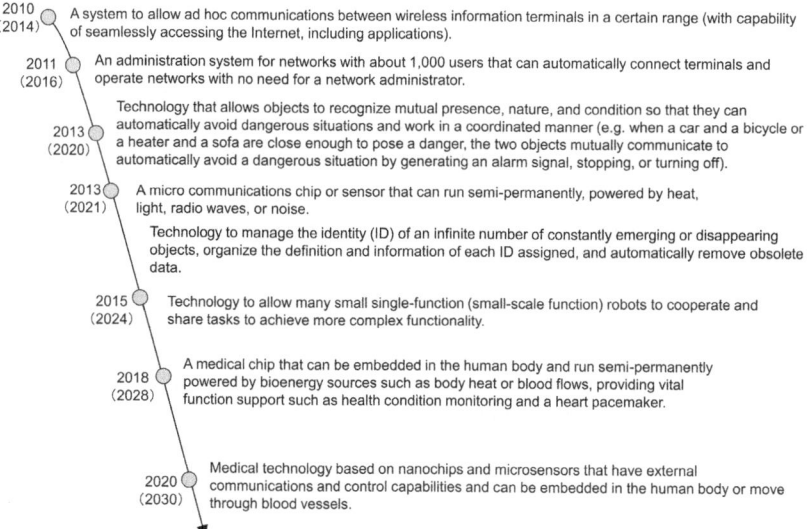

Fig. 2 Ubiquitous networking from the Delphi topics

stakeholders in various fields is obtained. The roadmap can predict technological difficulty and the global competitiveness of each technological element. However, it is difficult to measure these items empirically and quantitatively.

On the other hand, the Delphi method can give a quantitative knowledge of the realization time and the degree of importance of technologies, based on predictions by scientists and other experts. In this case, the accuracy of the results depends on the scale of the survey and the number of respondents. However, with Delphi results, it is difficult to grasp the connection of the technological topics or the future vision of the society that those technologies will create. This is easier with roadmapping. As one example, the results of a Delphi survey by the authors are shown in Figure 2. This figure is an arrangement of technological topics concerning a ubiquitous network in order of the forecast realization times of the technologies. Even though all these technological topics concern ubiquitous networks, it would be difficult to arrive at the image of the society that would be achieved if all the technologies were realized. Similarly, it is impossible to guess the connections between technological topics and the function of topics as turning points. Moreover, as described in the Introduction, there is a need to consider three aspects when conducting technology foresight in the future. That is, it is necessary

- To consider not only the realization of technologies, but also the social elements that require attention in the process of their realization.
- To bring as many stakeholders as possible into the process of technology foresight, because the necessity of a shared vision among stakeholders increases as technological uncertainties increase.

- To maintain high flexibility in the results of technology foresight as a management tool. Increased technological uncertainties imply an increase in the demand for after-the-fact revisions and modifications in technology foresight results.

Therefore, the following will examine integration of the Delphi method and technology roadmapping, considering the merits and demerits of both survey techniques and the above-mentioned three aspects (especially the third aspect).

5.2 Integration of Technology Roadmapping and the Delphi Method

As mentioned earlier, technology roadmapping includes a social vision or R&D concept, for example, the gradual miniaturization of semiconductor devices, development and spread of energy-saving fuel cells, and so on. The connections and relationships among individual technological topics, products, and markets are also visualized. However, roadmapping does not provide detailed information on each technology, such as the technological difficulties or R&D benchmarks in major countries.

This is the key point of one of the most beneficial aspects of integration of the Delphi method and roadmapping. That is, integration of the two techniques makes it possible to input various and quantitative detailed information from the Delphi data into roadmapping (see Figure 3). This means that informative contrasts can be obtained by multiplying the technological data from the Delphi survey in 'two-dimensional' roadmaps, thereby making the roadmaps more 'three-dimensional'. Because the Delphi data are an evaluation by scientists and other experts in the technological area concerned, this is extremely useful information for execution and achievement of the R&D strategy obtained by roadmapping. Moreover, these Delphi data may also become a factor in reviews of the roadmaps. The expert panel method, which is often used in roadmapping, is limited in terms of recognizing technological search areas. The Delphi results may even be able to influence the realization of the products and services at which the roadmaps aim.

As mentioned above, the recent increase in technological uncertainties makes it necessary to maintain high flexibility in the results of technology foresight, including roadmapping. In a set of Delphi topics, topics recorded in the roadmap coexist with those not recorded. These topics include technologies that exist in a rival relationship and never appear on the same roadmaps. Therefore, when correction of a roadmap is necessary for ex post facto reasons, substitution with other technologies such as rival technologies is possible (see Figure 4). In current roadmaps, unexpected ex post facto reasons might demand the revision of the whole roadmap, and this revision work may require a return to the Ver.1 roadmapping process. However, if the background of the roadmapping contains the Delphi method, the Ver.1 roadmaps would implicitly include rival technological topics as 'alternative options' when ex post factor revision becomes necessary. Thus, prior preparation of this 'alternative option' will function as a 'hedge' against the risk of unexpected ex post facto revisions.

In addition, a quantitative understanding of the strengths and weaknesses of a country on the roadmap is possible by referring to the Delphi results. Although this is difficult in the current roadmapping process, this is useful information. The technology roadmaps created by METI in Japan have a basic style, which is premised on the concept that all technology seeds and resources should be produced or secured domestically in Japan. However, with the increasingly advanced and complex technological systems of recent years, this kind of 'national self-sufficiency' has become rare, even in specific industries. Information on quantitative R&D benchmarks that shows the strengths and weaknesses of respective countries obtained by the Delphi method will facilitate thinking about cooperation with other countries.

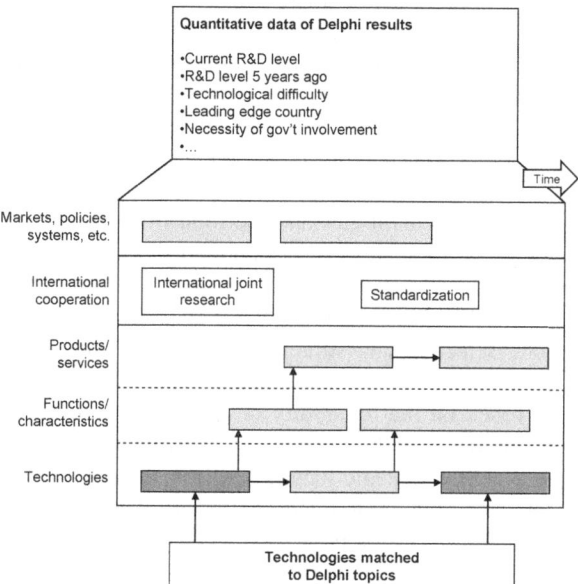

Fig. 3 Integration of the Delphi method and technology roadmapping (1)

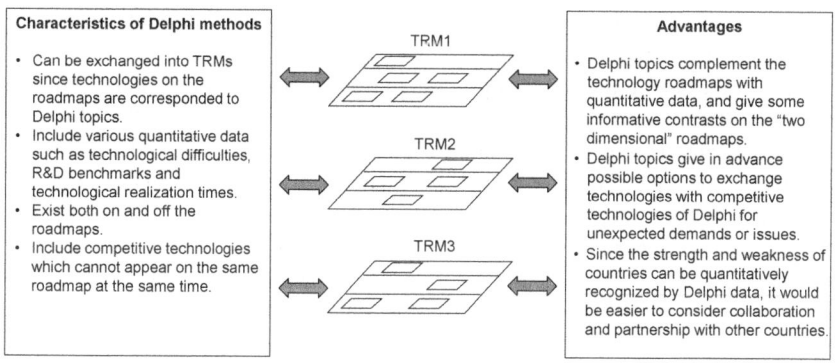

Fig. 4 Integration of the Delphi method and technology roadmapping (2)

5.3 Strategic Characteristics of Technology Foresight; Difference between Technology Roadmapping and Delphi Method

Finally, this section will discuss the relationship between technology foresight and the participation of various stakeholders. As mentioned earlier, bringing as many stakeholders as possible into the process of technology foresight is necessary because the necessity of a shared vision among stakeholders increases as technological uncertainties increase (Rip et al., 1995; Wilsdon and Willis, 2004). However, the authors' investigation implies that, if a larger number of stakeholders are brought into the technology foresight process, the foresight results tend to become all-inclusive (and indiscriminate) as the foresight process becomes more selective and strategic. On the other hand, when the technology foresight process uses a more inclusive and objective method, the foresight results will be more selective and strategic.

We will explain this again using an example of the Delphi method and technology roadmapping. Technology roadmapping begins with an R&D target, image of society, vision of the future, then gradually breaks down the issues and connects to the element technologies and the topics that should have the highest possibility of realization to achieve the technological realization time. The R&D strategy for reaching the final goal is also established.

However, as described here, in technology roadmapping, the panelists have already made quite selective and strategic judgments at this stage. That is, the technologies put on the roadmap are selected strategically from innumerable technology seeds based on the possibility of realization and other criteria. These technology seeds are identified by the panelists. Listing the technologies would be impossible without highly strategic judgments. Ironically, if various stakeholders, who may number in the hundreds including experts from industry, government, and universities, join the foresight process, the eventual roadmap is most likely to become extremely inclusive owing to the lack of a single coherent strategy. This is a natural result of the fact that stakeholders have different backgrounds and incentives concerning the roadmap. An attempt to narrow the range of these opinions will inevitably result in the 'greatest common factor'. At this point, the lack of a strategy becomes a problem, defeating the original purpose of establishing a strategy.

On the other hand, the forecast process in the Delphi method is both highly less subjective (less biased) and inclusive. It is no exaggeration to say that this is the most important feature of the Delphi method. First, scientists and experts in a certain specific field are identified. Next, each of these respondents evaluates the R&D benchmarks, technological importance, realization time, etc. anonymously and without weighting by affiliation or position. In the second round of the questionnaire, the respondents can refer to the other evaluations but cannot identify the respondents responsible for other answers.

This again excludes bias based on personality, position, and authority. Consequently, the results are very strategic and selective. Information on the technological realization time, degree of importance, and R&D benchmark data are listed by technological topic. Data which can be used as a strategic management tool is obtained, such as ranking data on the most important technological topics and Japan's strength in the technology. As a result, the selectiveness and strategic characteristics, which were excluded from the investigation design, are obtained in the results.

When designing integration of these two methods, sufficient consideration must be given to the purpose of technology foresight. Is the aim to contribute to policy making, or to build a strategy for an R&D consortium? The required results may not be obtained if technological forecasting is not based on a firm recognition of the purpose of the forecast.

6 Case Study of Integration of Integration of Technology Roadmapping and the Delphi Method: EU Nanoroadmap

As part of its Sixth Framework Programme (FP6), the European Commission (EC) created a roadmap for nanotechnology (the EU Nanoroadmap), and published it on a website in January 2006 (AIRI/Nanotec IT, 2006). The roadmap's purpose is to provide a medium- to long-term projection and outline for nanotechnology in three research fields (materials, health and medical systems, and energy) through 2015. This section introduces the EU Nanoroadmap as a case study of integration of technology roadmapping and the Delphi method.

6.1 Goal

Europeans engaged in R&D created the EU Nanoroadmap with the goal of providing knowledge in order to grasp the impact of nanotechnology on society and the economy and more effectively disseminate the results of R&D to the economy and society at large. Therefore, while the roadmap's users include managers and researchers in each sector, its messages for industry are particularly significant. It also emphasizes that small and medium businesses and venture firms are also targeted. The following are also goals of the roadmap:

- Strengthened international competitiveness and expanding markets in the nanotechnology field
- Improved selection, focus, and efficiency of R&D projects
- More effective training and education in the nanotechnology field
- Strengthened national and international collaboration in Europe
- Sustainable development and better quality of life in Europe.

6.2 Methodology

Creation of the roadmap took place over two years from 2004 to 2005 in the following two stages. The first stage was carried out during the initial year. It primarily involved the collection and analysis of information regarding nanotech policy and technology trends in various countries and sought to identify the fields where nanotech could be applied based on the results. The second stage involved the actual work of creating the roadmap. The results of each stage can be downloaded as reports from the project's website. In addition, international conferences were held each November to introduce survey results and gather the opinions of participants. An international consortium comprising technology consultants in different areas of expertise from eight EU countries and Israel was formed in order to carry out the survey.

The roadmap itself was created using the Delphi method. The number of Delphi panel respondents was about 230 (65 percent response rate). There were two question cycles, with the following main processes:

- Selection of leading international experts (Delphi panel)
- Creation of questionnaires for each technology field (including not only questions directly related to technology, but also many questions about examples of applications in society, the economy and industry, barriers to practical use, and technological benchmarks in various countries)
- Implementation of the first questionnaire (first cycle) using the Internet
- Collection of completed questionnaires and interviews conducted in relation to some of them
- Feeding back the results of the first cycle to the Delphi panel and implementation of the second questionnaire (second cycle)
- Creation of the final roadmap based on questionnaires, interviews, and international conferences.

6.3 Structure

The roadmap can be roughly divided into the following seven reports. Preliminary reports were created and published as Sectoral Reports for the three fields (materials, health and medical systems, energy), while technology roadmaps were created and published as the Synthesis Report for each of the three fields.

The technology roadmaps predict and analyze characteristics of various technologies as well as their advantages and disadvantages, and present their future applications over the coming 10 years. Preparation of the roadmaps centers on the applications of these technologies. The horizontal axis is the development phase (basic research, applied research, etc.) rather than time. The time axis concept is expressed in three maps for five-year periods beginning in 2005 (materials field). Subsequently, technological and social issues and bottlenecks are discussed. The international competitiveness of technology, accessibility of the integrated research infrastructure, need for integrated research facilities, and so on are broadly examined.

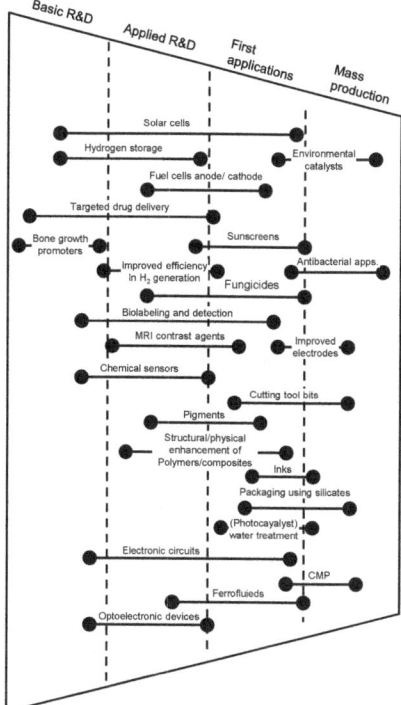

Fig. 5 Roadmap for the nanoparticle area in 2010 (Willems & van den Wildenberg, 2005)

The most characteristic feature of the roadmap is that the roadmap's horizontal axis represents four R&D phases (basic research, applied research, first applications, and mass production) rather than time. The time axis concept is expressed in three maps for five-year periods beginning in 2005, 2010 and 2015 (see Figure 5).

Figure 6 depicts estimated market growth over the next 10 years (vertical axis) and technological and economic risks accompanying R&D (horizontal axis) for feasible nanoparticle applications. Rather than risk per se, the horizontal axis can perhaps be thought of as depicting the "depth" of issues and "height" of barriers on the path to practical application. Returning to the example of solar cells, risk is moderate, while estimated market growth is highest of all.

As is shown in these figures, the data needed to build the roadmaps were effectively accumulated by the Delphi method. The technological characteristics such as R&D phase and Technological risk involved with R&D could not be obtained without experts' knowledge. This roadmap project also has examined the EU's international competitiveness by type of organization and necessity of establishment of multidisciplinary centers for nanotechnology industrial application based on the results of Delphi.

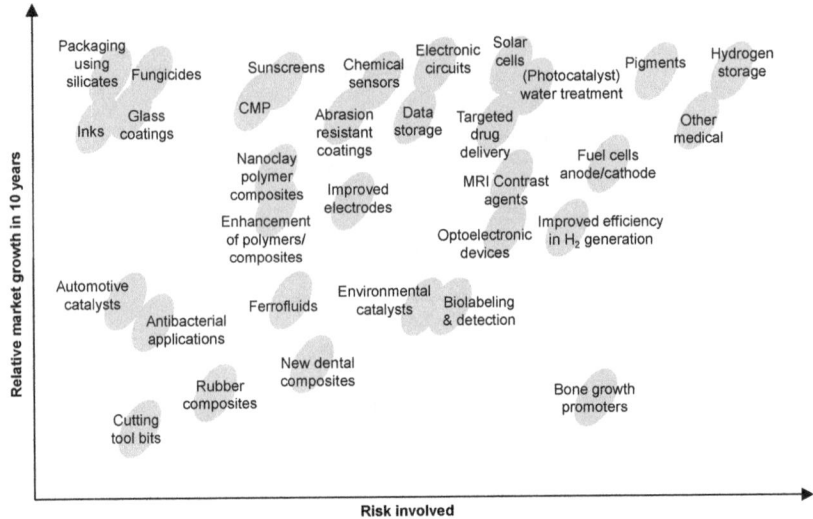

Fig. 6 Risk involved with R&D and estimated market growth for the nanoparticle area (Willems & van den Wildenberg, 2005)

7 Guideline on How to Manage Integration of Technology Roadmapping and the Delphi Method

Ideally, integration of technology roadmapping and the Delphi method is thought to be very effective. However, the actual process must be a based on a plan that can be executed naturally. One example of an action plan for integration of technology roadmapping and the Delphi method is shown in Figure 7. It is necessary to complete all this work within a limited time of a few years to avoid the influence of changes in the social environment, political needs, or remarkable technological development.

First, a technology foresight committee is necessary to organize the survey as a whole. This committee should lay out alternative visions of the future society on which the survey will be based. Not only technological experts, but also policymakers, managers, and journalists must be encouraged to join. Technology roadmapping expert panels should be held based on the visions of the future society and short scenarios drawn up by the committee. An outline of the technology roadmap is built based on each vision. Respective roadmap working groups take over these outlines and construct detailed sub-roadmaps.

The Delphi panels are held simultaneously with the roadmapping work. It is necessary to identify the technological topics comprehensively in each technological field. Therefore, the Delphi panels should be held by technological

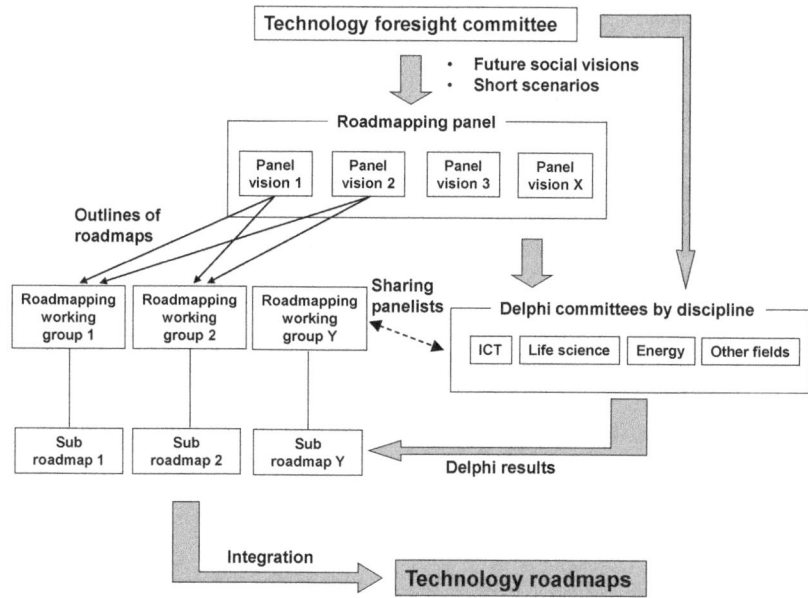

Fig. 7 Guideline of an integrated project using the Delphi method and technology roadmapping

field, such as information and communication technologies and life sciences. As a standard, fewer than 100 technological topics are identified by each Delphi panel, resulting in a total of several 100 topics. These topics and the results of the Delphi survey are given to the roadmap working groups at any time and are reflected in the roadmaps. It is preferable that each panelist participates in both the process of the roadmap working groups and the Delphi process to improve efficiency in exchanges of information between the two processes.

Each technology roadmap, which consists of Delphi results, is eventually integrated at the technology foresight committee. These are used to establish R&D strategies or in investment planning. The integrated technology roadmaps properly reflect the progress of changing social conditions and technological development, and should be revised accordingly.

Many people in this field may already have experienced with both processes. Nevertheless, execution and integration of both surveys will be a major challenge, since the Delphi surveys are implemented in various conventional technological fields, while roadmapping is implemented based on a vision of the future society. Whether information with these different characteristics can be shared effectively in the short term, and how the survey system is to be constructed, will be key points in this challenge, even though the panelists participate in both processes to ensure efficient sharing of information.

8 Conclusions

The interaction between society and science and technology, and between industrial competitiveness and Science and technology, is becoming closer, while increased technological complexity and sophistication are increasing technological uncertainties. Given this situation, technology foresight is no longer a simple forecast of future trends in technology, but also includes the establishment of a vision of the future society that the government, industry, and society hope for and aim to achieve. From this viewpoint, technology foresight must include the following three aspects:

1. Technology foresight must consider and include not only technological topics, but also social issues which must be solved in the application of those technologies.
2. It must be possible for various stakeholders to participate in the foresight process.
3. High flexibility must be maintained in the foresight results when using technology foresight as a management tool.

In particular, finding a solution for 3 is a challenge for the research community. The establishment and proposal of a more efficient and effective R&D management tool to decision-makers will have enormous significance for Science and technology policy.

The Delphi method was originally devised by a US think tank. However, it has been developed mostly in Japan, where the linkage between the Delphi method and policy-making has been especially strong, being used to determine the priority research areas in the country's Third Science and Technology Basic Plan. On the other hand, technology roadmapping is one of the most developed R&D management tools in the world. Today, technology roadmaps are likely to imply a social vision or concept, based on which connections of technological topics, products, and markets are visualized. Integration of the Delphi method into technology roadmapping provides quantitative Delphi data for 'two-dimensional' roadmaps, and thereby makes them 'three-dimensional'. These 'information contrasts' can have an obvious benefit.

Implementation of integrated technology foresight is a difficult challenge. In this chapter, the authors proposed a possible action plan. Numerous problems must be solved considering the limited time and human resources available for implementation. A repeated trial-and-error approach to the detailed process and appropriate modification of the plan are necessary. Finally, both strong internal and international collaboration are required.

References

AIRI/Nanotec IT, Roadmaps at 2015 on nanotechnology application in the sectors of materials, health & medical systems, energy. Synthesis report, EU Nanoroadmap Project, EC 6th Framework Programme. VDI/VDE Innovation und Technik GmbH, Berlin (2006)

Banuls, V.A., Salmeron, J.L.: A scenario-based assessment model - SBAM. Technological Forecast and Social Change 74(6), 750–762 (2007)

Caldwell, B., Wang, E., Ghosh, S., Kim, C., Rayalu, R.: Forecasting multiple generations of technology evolution: challenges and possible solutions. International Journal of Technology Intelligence and Planning 1(2), 131–149 (2005)

Chesbrough, H.W.: Open Innovation: The New Imperative for Creating and Profiting from Technology. Harvard Business School Press, Cambridge (2003)

Eto, H.: The suitability of technology forecasting/foresight methods for decision systems and strategy – a Japanese view. Technological Forecast and Social Change 70(3), 231–249 (2003)

Goto, A., Kodama, T.: Japan's National Innovation System: Rebuilding the Engine of Growth. University of Tokyo Press, Tokyo (2006)

Goto, A., Odagiri, H.: The Industrial Systems in Japan and New Developments 3 - Science-based Industries. NTT Publishing Co., Ltd., Tokyo (2003)

Kameoka, A., Yokoo, Y., Kuwahara, T.: A challenge of integrating technology foresight and assessment in industrial strategy development and policymaking. Technological Forecast and Social Change 71(6), 579–598 (2004)

Kanama, D.: EU Nanoroadmap - issues and outlook for technological roadmaps in the nanotechnology field. Science and Technology Trends (23), 55–64 (2007)

Karube, I.: Integration of socio-economic needs into technology foresight. In: Proceedings of the International Conference on Technology Foresight, Report Material No. 77 (2001)

Kondo, M.: University-industry partnerships in Japan. In: Nagaoka, S., Kondo, M., Flamm, K., Wessner, C. (eds.) 21st Century Innovation Systems for Japan and the United States: Lessons from a Decade of Change: Report of a Symposium, pp. 186–205. The national academies press, Washington, DC (2009)

Kuwahara, T.: Technology forecasting activities in Japan. Technological Forecast and Social Change 60(1), 5–14 (1999)

Kuwahara, T.: Technology foresight in Japan - the potential and implications of the Delphi approach. In: Proceedings of the International Conference on Technology Foresight, Report Material No. 77 (2001)

Landeta, J.: Current validity of the Delphi method in social science. Technological Forecast and Social Change 73(5), 467–482 (2006)

Martin, B.: Technology foresight in a rapidly globalising economy. A Paper Prepared for The UNIDO Regional Conference on Technology Foresight for Central and Eastern Europe and the Newly Independent States, UNIDO, Vienna (2001)

Martin, R.: Technology roadmaps: infrastructure for innovation. Technological Forecast and Social Change 71(1), 67–80 (2004)

NISTEP, Outlook for Japanese and German Future Technology - Comparing Japanese and German Technology Forecast Survey. NISTEP Report No. 33. National Institute of Science and Technology Policy, Tokyo (1994)

NISTEP, The Seventh Technology Foresight - Future Technology in Japan Towards the Year, NISTEP Report No. 71. National Institute of Science and Technology Policy, Tokyo (2001)

NISTEP, The 8th Science and Technology Foresight Survey - Delphi Analysis. NISTEP Report No. 97. National Institute of Science and Technology Policy, Tokyo (2005a)

NISTEP, Comprehensive Analysis of Science and Technology Benchmarking and Foresight. NISTEP Report No. 99. National Institute of Science and Technology Policy, Tokyo (2005b)

NISTEP, Social Vision Towards, - Scenario Discussion based on Science and technology Foresight. NISTEP Report No. 101. National Institute of Science and Technology Policy, Tokyo (2007)

NISTEP, The 9th Science and Technology Foresight - Contribution of Science and Technology to Future Society - Future Scenarios Opened up by Science and Technology. NISTEP Report No. 141. National Institute of Science and Technology Policy, Tokyo (2010a)

NISTEP, The 9th Science and Technology Foresight - Contribution of Science and Technology to Future Society - The 9th Delphi Survey. NISTEP Report No. 140. National Institute of Science and Technology Policy, Tokyo (2010b)

Petrick, I.J., Provance, M.: Roadmapping as a mitigator of uncertainty in strategic technology choice. International Journal of Technology Intelligence and Planning 1(2), 171–184 (2005)

Phaal, R., Farrukh, C.J.P., Probert, D.R.: Technology roadmapping - a planning framework for evolution and revolution. Technological Forecast and Social Change 71(1), 5–26 (2004)

Phaal, R., Farrukh, C.J.P., Probert, D.R.: Developing a technology roadmapping system. In: Proceedings of PICMET 2005, pp. 99–111 (2005)

Preez, G.T.D., Pistorious, C.W.I.: Technological threat and opportunity assessment. Technological Forecast and Social Change 61(3), 215–234 (1999)

Rip, A., Misa, T.J., Schot, J.: Constructive technology assessment: a new paradigm for managing technology in society. In: Rip, A., Misa, T.J., Schot, J. (eds.) Managing Technology in Society: The Approach of Constructive Technology Assessment, pp. 1–14. Pinter Publishers, London (1995)

Stokes, D.E.: Pasteur's Quadrant: Basic Science and Technological Innovation. Brookings Institution Press, Washington, DC (1997)

UNIDO, UNIDO Technology Foresight Manual, vol. 1. United Nations Industrial Development Organization, Vienna (2005)

Willem, van den Wildenberg: Roadmaps at 2015 on nanotechnology application in the sectors of materials, health & medical systems, energy: Roadmap Report on Nanoparticles. EU Nanoroadmap Project, EC 6th Framework Programme (2005)

Willyard, C.H., McClees, C.W.: Motorola's technology roadmap process. Research Management 30(5), 13–19 (1987)

Wilsdon, J., Willis, R.: See-through Science: Why Public Engagement Needs to Move Upstream. Demos, London (2004)

Wounderberg, F.: An evaluation of Delphi. Technological Forecast and Social Change 40(2), 131–150 (1991)

Yasunaga, Y., Yoon, T.: Technology Roadmaps: Creation of New Industries through Overviews and Analysis of Technological Knowledge. Open Knowledge Corporation, Tokyo (2006) (in Japanese)

Author

 Since April 2010, **Daisuke Kanama** has been an Associate Professor of the Department of Business and Information Systems at the Hokkaido Information University, and a Visiting Research Fellow at the Technology Foresight Center of the National Institute of Science and Technology Policy (NISTEP), Ministry of Education, Culture, Sports, Science and Technology (MEXT). His research focuses on application of science and technology foresight methodologies in innovation strategy planning. He also carries out research on the university–industry collaboration and intellectual properties.

The Innovation Support Technology (IST) Approach: Integrating Business Modeling and Roadmapping Methods

Hitoshi Abe

The purpose of this paper is to report the integration studies of business modeling and roadmapping methods for the "Innovation Support Technology" (IST) and the Innovation Support Technology's practical application to real-world cases. The Innovation Support Technology is conducted for the purpose of offering a convenient tool for engineers and researchers in order to enhance corporate value from R&D outputs. "Japan's lost decade" has forced companies to change R&D management and R&D operation style, especially regional industries. We propose the framework for revitalization of regional industries by using the strategic technology roadmap made by the Ministry of Economy, Trade and Industry (METI-TRM) with business modeling. We applied this innovation support technology method to several real-world cases to show its effectiveness.

1 Introduction

Challenges to solve social problems such as environmental issues, the falling birthrate and the aging population are continuously demanding. Moreover, product life cycles have become shorter and shorter by globalization and acceleration of technological innovation. The speed of value migration and the degradation of existing product values are accelerating. Product innovation becomes more important for companies, especially R&D driven companies, to maintain and accelerate their growth.

As R&D driven companies are adapted for rapidly changing economical circumstances, they need to clarify what to make and to accelerate to create business value from R&D outputs. With progress in open innovation, engineers and researchers need to communicate more and more frequently with people not only in-house but also outside the company and the investors in order to create business value from R&D outputs.

Hitoshi Abe
Japan Techno-Economics Society 3-3-1,
Iidabashi, Chiyoda-ku, Tokyo 102-0072,
Japan
e-mail: abe@jates.or.jp, abeh@muc.biglobe.ne.jp

For fulfilling those needs, we have started a study on the business modeling method in a group at JATES[1] (Japan Techno-Economics Society) since autumn in 2002. The business modeling method for engineers and researchers has been theoretically studied, developed to solve such problems, and applied to the real-world cases. Results from these case studies were fed back to the modeling theory and the implementation of the modeling processes. An easy-to-use business modeling method for practitioners has been developed (Abe et al., 2004; Abe et al., 2005; Ishida et al. 2006).

In order to form a framework of the business model for our study, we mainly referred to the studies by Chesbrough (2003), Hamel (2000), Slywotzky (1998, 1997, 2002), MacMillan (1995) and Ikeda and Imaeda (2002).

After Slywotzky, a business model (BM) is described as follows: *the totality of how a company selects its customers defines and differentiates its offerings, defines the tasks it will perform itself and those it will outsource, configures its resources, goes to market, creates utility for customers, and captures profit. It is the entire system for delivering utility to customers and earning a profit from that activity.* We use, on the one hand, business model to support engineers to envisage "what", "who" and "how" are conditioned for their innovation, and to draw "how much" cost and value are required for its financial model.

On the other hand, we used strategic roadmapping (SRM) enabling the various functions and perspectives to be aligned. Strategic roadmapping is a well known and commoditized method that comprises a time-based, multi-layered chart.

In preliminary studies conducted in 2006 (Abe et al., 2006) we analyzed a new strategic business planning method for integrating the BM and the strategic roadmapping by supplementing the weak points of the business modeling method (see Table 1) with the strategic roadmapping method (Bucher, 2002; Kameoka, 2003; Phaal et al., 2005; Tschirky et al., 2003).

Table 1 Comparison of business modeling (BM) and strategic roadmapping (SRM)

Pros of Business Modeling (BM)	Cons of Business Modeling (BM)
1. To know how to create company value from R&D Outputs and provide an operation model. 2. Modeling tool to create Business Concept from Business Idea 3. Modeling of the competitive strategy: how, what, or to whom you provide the Service/product, how to win competition.	1. Difficult to find out market trends and opportunities 2. Difficult to make a decision of investment timing 3. Difficult to judge the choice of an alternative technology 4. Difficult to know by when & what technology should be developed
Pros of Strategic Roadmapping (SRM)	**Cons of Strategic Roadmapping (SRM)**
1. Roadmap consists of layers, such as market, business, products, technology and resources, that are systematically expressed on a time-axis and provide a landscape. 2. It can be utilized as a strategy planning tool, which supports and opportunities, choice of an alternative technology, and associates the elements between layers. 3. Knowledge creation for a better action: discovery of gap, discovery of bottlenecks, discovery of technological defects and promotion of development, estimate of required resources.	1. Difficult to evaluate business value 2. Difficult to express a business attractiveness of R&D outputs 3. Difficult to express a business system or operation model 4. It takes more time to create and maintain Roadmap under satisfying comprehensiveness

---

[1] JATES is a public-interest membership society mainly for companies involved in research, development and innovation in support of their business activities. IRI and EIRMA are counterparts of JATES in the U.S. and in EU, respectively.

The purpose of this paper is to report about the integration studies of the business model and the roadmapping methods for the Innovation Support Technology (IST) and also about the practical application to the real-world cases of the latter. We propose the framework for revitalization of regional industries and medium-size companies by using the Strategic Technology Roadmap made by the Ministry of Economy, Trade and Industry (METI-TRM) (2006) with the Innovation Support Technology (Abe et al., 2007).

2 Concept Value and Utility of the Innovation Support Technology

The concept of the Innovation Support Technology which is based on the best use of the pros of business model and strategic roadmapping is present in this chapter.

2.1 Concept of the Innovation Support Technology

By integrating business model and strategic roadmapping, the Innovation Support Technology is a methodology that manages to integrate the two concepts of technology push and market pull in one and the same approach. While business model is a tool that focuses on technology push, strategic roadmapping covers the concern of market pull.

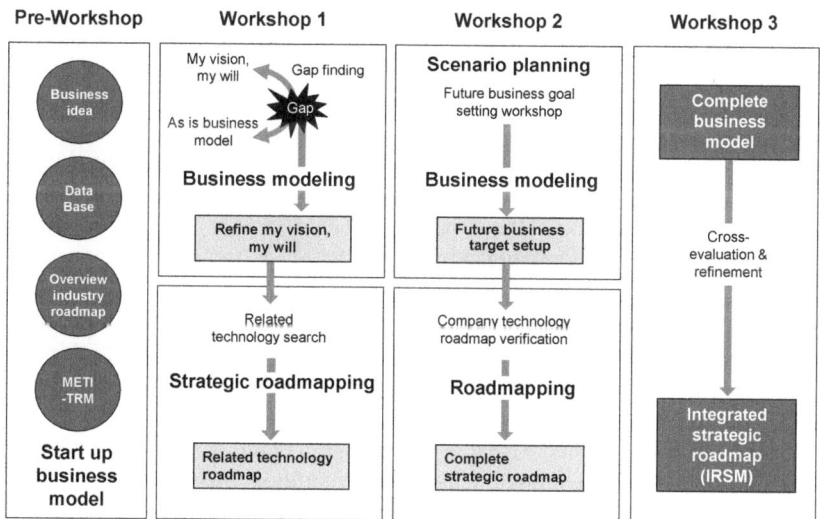

Fig. 1 Framework and work procedure of the Innovation Support Technology

Thus, Innovation Support Technology is a methodology that promotes corporate value creation from R&D outputs, without leaving out of sight the market perspective of each technology-based innovation. In this study strategic roadmapping and business model methods are further supported by METI-TRM (the METI-Strategic Technology Roadmap). The implementations scheme of the Innovation Support Technology is shown in Figure 1.

The Innovation Support Technology method can offer quicker decision-making, better exploration of innovation opportunities as well as better business plan than those by the conventional methods, because the Innovation Support Technology method multilaterally examines the plan by both time and resource axes. Thus, the operation plan can be firmly established by the Innovation Support Technology method.

Furthermore, one can develop technology development and product development plans based on social trends and the resource situation, and it becomes easy to make adjustments on the time axis in investment timings.

Moreover, the followings are obtained by using METI-TRM as a technological knowledge data base that contains the mass of specialists' wisdom:

1. The proof of technological trends is executed, and the business scheme becomes more visible;
2. Explore the possibility of new product/market development based on the understandings of technology trends and different technical fields.

2.2 *Value and Utility of the Innovation Support Technology*

The value and utility of the Innovation Support Technology is to create customer value and to solve customer problems, which are achieved by the planning process itself and the outcomes, namely the integrated strategic roadmap (ISRM). The strategic planning procedure by the Innovation Support Technology is important because the quality of the business plan and new product planning depends on the process. The milestone management of business, product, function, technology and resources could be made by the integrated strategic roadmap.

The business model and the roadmap created by the Innovation Support Technology are used as communication tools that offer a common language and share the whole image of a strategy and a new business with people of different standpoints. Recognition and understanding is deepened in the Innovation Support Technology design process. Detecting of a gap and resetting of the problem are made visible by an arranged simple diagram. The quality of the solution and the decision-making can be improved as a result.

The use of the Innovation Support Technology lets the participants understand the future from what tried at random together. In addition, by using business model and strategic roadmapping, one can see what one didn't see before.

Business model and strategic roadmapping can divide possibility and risks and show how we should act. Furthermore, in the case of a new business, we extract an uncertain factor by the business model and strategic roadmapping work. When one doesn't well understand what is correct, one can use the integrated strategic roadmap as a trial plan to make unclear and ambiguous things visible.

2.3 Benefits of the Concept

The Innovation Support Technology method can provide quicker decision-making, better exploration of innovation opportunities as well as better business plan than those by the conventional methods, because the Innovation Support Technology method multilaterally examines the plan by both time and resource axes. Thus, the operation plan can be firmly established by the Innovation Support Technology method.

2.4 Profiling Specific Approaches

The Innovation Support Technology method is useful not only for R&D driven companies but solution and service oriented companies in technology-intensive sectors. In organizational size and structure, however, the Innovation Support Technology method can be more suitably applied for the development team and division rather than for corporate level, and for business planning in business unit and project planning in R&D level rather than for communication tool in corporate level. In business life cycle, this method demonstrated the real powerful planning tool for the second start-ups of the medium size corporation and the division of the large scale company.

The Innovation Support Technology is specially recommended as a tool to plan and discuss and finally determine the new business target related with multi-business divisions of the big company and/or related with plural medium size companies, namely many stakeholders involved. Planning should be down by the cross functional team.

3 Framework and Work Procedure of the Innovation Support Technology

Framework and workshop procedure of the Innovation Support Technology are schematically shown in Figure 1. The planning of the Innovation Support Technology starts the pre-workshop before conducting workshops which are divided into three steps. The workshop is carried out in group work style.

In the pre-workshop, the leader should design a guidance program including the purpose, the goal, working image and necessary knowledge. It is recommended that the leader is empowered by the top management to select participants from cross functional fields and control assistant staffs to prepare required data base and support the workshop.

3.1 Workshop 1

The product concept and the business idea based on R&D outputs are described. Based on this product concept and business idea, the market and the customers are pictured and determined. The business idea based on an existing its own business system is expressed by the business model framework as an "as is model". A business concept as "to be model" is led and expressed by a corporate strategy, a business strategy or "my vision & my will" from researchers and engineer's desires. The gap between the "as is model" and the "to be model" should be found and clarified in phase 1. Then, reasons for the gap and the solution approach are discussed and oriented. The next working procedure is to make a technological scenario. Product functions and technologies related to the product concept drawn with "my vision & my will" are pulled out from the METI-Strategic Technology Roadmap (METI-TRM) database and arranged for the purpose, and then a technological scenario is made. By this method the technological scenario can be easily obtained by use of the METI-TRM database, not by zero-base.

3.2 Workshop 2

Next, planning of the business scenario is made. Industrial value chain analysis, PEST (political, economic, sociological, technological) analysis, and business environment analysis by using five forces after M. Porter are executed as a start of the business modeling procedures (Porter, 1980, 1985, 1998). Then we create profit modeling by utilizing Slywotzky's 22 "profit patterns" in business activities (Slywotzki, 1998, 1997, 2002).

The work step of the business scenario plan is detailed in Ikeda and Imaeda (2002), Abe et al. (2006), Heijden (1996) and Robert and Smith (1990). Scenario drivers are extracted from PEST analysis as a function of influence and uncertainty to its company. The business concept and the product concept obtained from "my vision & my will" are carefully expressed and designed by the business modeling method to make the target clearer in the future.

1. The purpose of the business scenario plan is to design and to obtain the business target of the company in the future.
2. The company's business unit technology roadmap can be obtained by roadmapping of the product function and the enabling technology to achieve the business target in the future.
3. This work is the key of the Innovation Support Technology work process.

3.3 Workshop 3

Contents of the business unit technology roadmap and discoveries through these roadmapping workshops are reflected in the business model in the future. The target customers, the value propositions, the supply method, and the profit model according to scenarios are confirmed before the business model is completed. Then, the business model and the company technology roadmap are integrated into the integrated strategic roadmap. Milestone gaps between layers, discoveries such as bottlenecks and the investment timing are investigated and verified by this integrated strategic roadmap. The business strategy is also evaluated. Evaluation results are fed back to the start-up business model if there is an imperfect part. Procedures of the Innovation Support Technology are repeated until the business model and the company technology roadmap reach the required level. The business model, the business unit technology roadmap and the integrated strategic roadmap are completed.

3.4 Innovation Planning Process by the Innovation Support Technology in Practice

The Innovation Support Technology meeting will be executed in every two weeks dividing into five times. The workshops are held three times from the 2nd to the 4th.

At the first meeting, a guidance program is introduced to participants by the program leader. The purpose is shared, the goal is confirmed, a work image and necessary knowledge are confirmed, and the promotion system of work is decided. At the second meeting (workshop 1), "my vision & my will" is presented by the leader of this workshop. All participants join in a group to examine the business model to find gaps between the "as is model" and "to be model" and the related technology roadmaps to obtain a technology scenario. At the third meeting (workshop 2), the improved business model and its business unit technology roadmap are developed as a group project. At the fourth meeting (workshop 3), the business model, the business unit technology roadmap and the integrated strategic roadmap are completed. At the fifth meeting, outputs of the workshop are reported by the leader to the stakeholders of the business strategy. Comments are fed back to the integrated strategic roadmap.

4 Case Study

In this chapter, we report a real-world case study, that is, the second start-up business plan for welfare service business in Ikeno Tsuken Co., Ltd.

4.1 Outline of Welfare Service Business in Ikeno Tsuken Co., Ltd.

Ikeno Tsuken Co., Ltd. is doing telecom-related business, electric-related business, engineering enterprise, solution business, and welfare service business. Its total sales amount in fiscal year 2005 is 15.6 billion yen, and that of its welfare service business is 400 million yen. The welfare service business provides visually impaired people walking support equipment/system, and construction / maintenance for the system.

There are two types of the support system, the stick method and the radio-wave method (see Figure 2):

1. The system of the *stick method* is outcome of the R&D which NEC started in 1982 for the purpose of effective utilization of by-product ferrite from industrial goods, sponsored by JST. When exclusive use stick comes near to magnetic marker zone guide path, top-end of the stick vibrates and informs where the guide path is. When visually-impaired person passes through on fork in a road or room entrance, ground-buried antenna catches movement of exclusive use stick and voice guide comes on the air. This system is composed of installed "voice guide equipment" and "exclusive use white stick" carried by a user. The system supports safety (other disabled people are also considered) and ease of mind (specific information about the destination is obtained).

2. This system of the *radio wave method* leads the way for a visually-impaired person, using voice emitted from a target which catches weak radio wave from a signal aid. This system is composed of installed "voice signal guide equipment" and a "signal aid" carried by a user and he/she can get voice guide when and where needed. This system provides information about location (direction to the target), voice loudness (distance information to the target) and contents (specific information about the target, to a visually-impaired person, since the target itself emits the voice).

**Stick method
(magnetic signal guide system)**

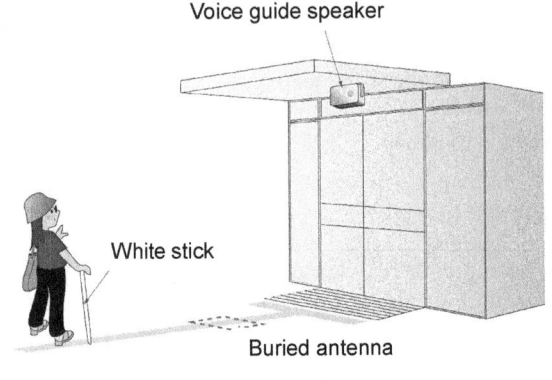

**Radio-wave method
(voice signal guide system)**

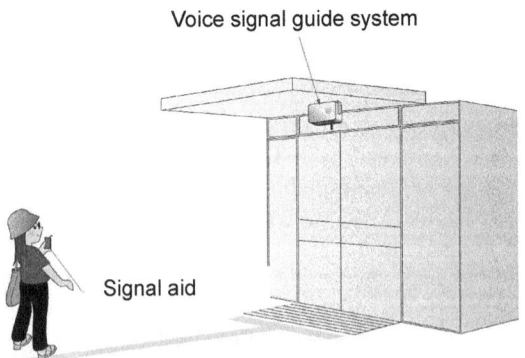

Fig. 2 Outline of welfare service business of Ikeno Tsuken Co., Ltd.

Table 2 The business modeling results

		B to G	B to B (to G)	B to G to C
Market &customer		• Tokyo Metropolis • Saitama City • Tachikawa City, et. al	• Kan electric construction • Taisei Construction, et. al	• Visually-impaired people (300,000) • Elderly people, et. al. • Total disabled people (3.2 million)
Products &services supplied		• Voice signal system (@1million yen) • Felt with fingers guide system (@3 million yen)	• Same as the left	• Terminal (@\7,000〜\12,000) • Sales performance→ 20,000 accumulation
Enterprise operation system		Proposal → design → order entry → delivery → inspection → maintenance fables	1. Catalog order items, standard items 2. Special order items	• Personal buys • Government buy and provide to personal
Profit model		• Equipment sales • Maintenance	• Equipment sales	• Equipment sales (Terminal)
Growth model	2006now 5 years later 10 years later 15 years later	350 million yen 350 million yen 350 million yen 600 million yen	4.7 million yen 150 million yen 200 million yen 300 million yen	3 million yen 5 million yen 10 million yen 100 million yen

Gap Measures

\4 Billion Business

4.2 Workshop 1

"My vision & my will" based on corporate managerial strategy of Ikeno Tsuken Co., Ltd. is as follows: "From visually impaired person walking support system to community support system. Make it to 10 times bigger business (4 billion yen scale) in 10 years".

In our initial business model, 10 years later sales amount of B to G (Business to Government) business is predicted 350 million yen, that of B to B (to G) is predicted 200 million yen, and that of B to G to C (Consumer) is predicted 10million yen (see Table 2). As a consequence, sales amount in 10 years will become 560 million yen in total. Therefore, we faced a big gap between "my vision & my will" of corporate managerial strategy and predicted results that were done mainly by the person in charge of this business.

Then, we invited an executive of the company to join our examinations, and searched new business opportunities together to fill the sales amount gap. We began with developing technology scenarios, referring to the METI-Strategic Technology Roadmap (METI-TRM). From "usability-related roadmap", we found out that sensor technology, communication technology and recognition-related technology will be developed by 2012, and they will be put into practical use. In the case of the technology for this business, for example, the people sensor and environmental sensor will be completed by 2012.

4.3 Workshop 2

On the basis of the technology scenarios extracted from the METI-Strategic Technology Roadmap (METI-TRM), industrial value chain analysis, PEST analysis, and Michael Porter's five forces analysis were executed. The factors from PEST analysis are as follows:

1. Political factors: Legal revision (barrier-free new law, etc.).
2. Economic factors: Changes of population profile (the aging of the population), changes of social consciousness (growing concern about the barrier-free environment, universal design, and safety and security).
3. Sociological factors: The trend of national projects such as city planning.
4. Technological factors: Appearance of the new Internet based systems, ubiquitous and nanotechnology.

Then, we moved on to scenario planning. As scenario driver candidates that have both a large influence on the company and are of high uncertainty, we listed

1. emergence of a new system that has superior spec than the existing system,
2. technology innovations such as a new system using the Internet, ubiquitous-technology or nano-technology,
3. changes in social awareness about the barrier-free environment and universal design.

In the next step, we picked up "universal design needs" and "technological capabilities allowing social participation" as scenario drivers and depicted four scenarios (see Figure 3). As the scenario that aims at expansion of the business scale, livelihood support systems for elderly people and disabled people were discussed. They include

1. magnetic signal guide system for hearing-impaired people,
2. emergency evacuation information system,
3. emergency report system, etc.

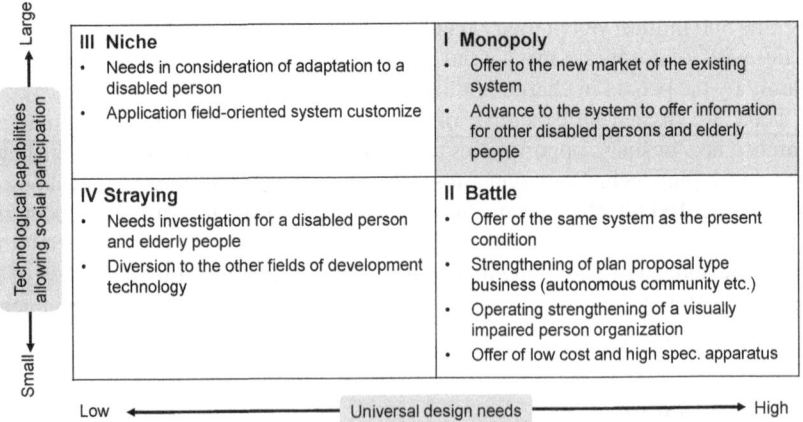

Fig. 3 Finding out the scenarios

Table 3 Company technology roadmap

		2005	2010	2015
Product	• New product/existing market • Community support system			• Super safety lamp (100units@10M yen) • Emergency evacuation system • (100Kunits@10Kyen) 2 billion yen
Market	• New market/existing product • Walking support system for impaired	• Extraction of term for standardization • Direction for standardization of each terms (test&proof)	• Consideration for utilization • Open specification of system • Module concerning construction	• For the elderly, and other impaired person • Relate to traffic infrastructure signal • From character recognition to voice recognition system • About 2 billion market ∗ Universal design
Function	Function factor • Terminal	• Standalone terminal type • Mobile phone built-in type • Compatibility with existing system • Light weight • Low cost	• Longer communication distance • Durability • Light weight • Low cost • Ubiquitous • Networking	• Multi-function type • GPS-function type • Light weight type • Low cost • Long time usability • Private information available
Technology	Technological factor • Sensor network • GPS • Infrastructure	• FM radio wave (weak) • Infra-red light • Visible light • Sensor	• Addition of RFID to infrastructure • IP protocol • Ubiquitous network	• IPv6 available • P to P Protocol standard • Mobile fuel cell • Light weight flexible display • Sensor communication technology

Table 3 shows the company technology roadmap from now to 2015. The sales amount for the community support system market (products such as emergency report system and so forth) will be an estimated 2 billion yen and that of the walking support system market (products such as elderly people walking support systems and so forth) will be estimated also at 2 billion yen in 2015.

4.4 Workshop 3

"The business model (10 years later)" was completed (see Table 4). Market and products predicted in 10 years are

1. visually impaired person walking support system /sales amount=550 million yen,
2. disabled people and elderly people walking support system /sales amount=2 billion yen, and
3. community support system /sales amount=2 billion yen.

That is, we successfully developed the business execution plan that will realize a 4.55 billion yen total sales amount in 2015. Based on the business model (ten years later), the business unit technology roadmap was reconstructed to the multi-layered strategic roadmap. "The integrated strategic roadmap" was completed (see Figure 4).

Table 4 The business model (10 years later)

Biz-model / Products & service			Basic business		
			• Visually impaired walking support system	• Walking support system for impaired & elderly person	• Community support system
Static	Strategic model	Domain	Existing products/existing market • Local government • General contractor/subcontractor • Railway carriers etc.	Existing products/new market • National police agency • Local government • Railway carriers etc. • Hotel/department store	New products/existing market • Ministry of Land, Infrastructure and Transport • Local government • Shopping precinct • Other private demand
		Value proposition	• Visually impaired walking support system • Announcing function at public facilities etc. by signal aid.	• Walking support system for impaired & elderly person • Signal aid • Announcing function at intersection • Elongation time during green traffic light	• Safety community support system (including risk management) • Building-up system by adjusting. • Substitution function of police box
	Supply method		• Planning proposal/direct sales	• Planning proposal/direct sales	• Planning proposal/direct sales
	Profit model		• System sales • Equipment sales • Maintenance • Sales 550M yen	• System Sales • Equipment sales • Maintenance • Sales 2 billion yen	• System sales • Maintenance • Sales 2billion yen
Dynamic	Sustainability business growth	Influence	• Acceleration of penetration by certification of signal aid as daily necessities • Providing equipment to railway carriers under traffic barrier-free law	• For other impaired (Utilizing signal aid) • Development and providing terminal	• Urban planning with safety, high quality, barrier-free concept (providing software & hardware)

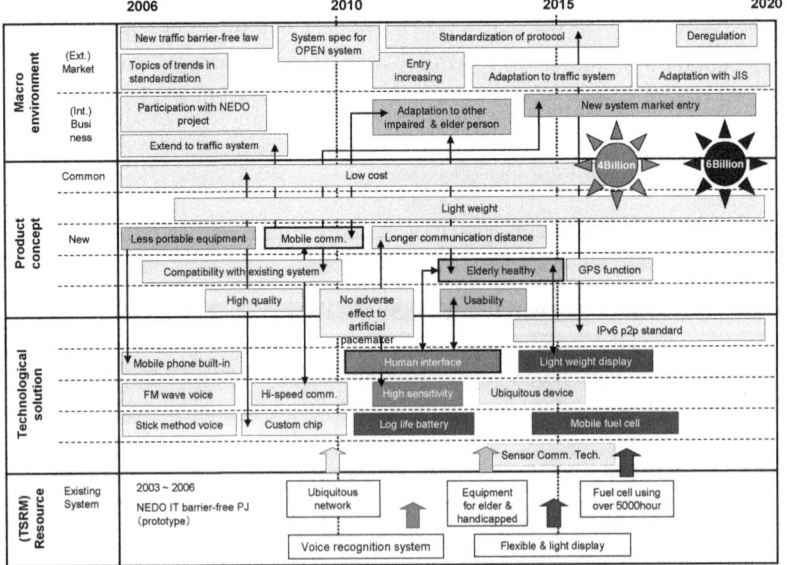

Fig. 4 The integrated strategic roadmap for a visually impaired walking support system

4.5 Observations about the Practical Applicability of Innovation Support Technology

What problems we observed during preparation and implementation in each of our three workshop steps and what critical success factors we derived out of these problems for each of the steps are just to describe "my vision & my will" first of all. "My vision & my will" was hardly derived by the workshop leader alone. Careful support and advice by the program leader is essential for the workshop leader to make a clear concept of "my vision & my will". Once "my vision & my will" was shown, business modeling and roadmapping are routine tasks rather than challenging work. The program leader is expected to be not only a technology management professional and technology road-mapping expert, but have skill sets of business consultant.

Critical success factors are pointed out as follows: The number of participants is from five to seven. Cross-functional participants are strongly recommended like marketing, R&D, sales, corporate staff, etc. The career level of people is less sensitive. However, lower level managers are recommended for a challenging theme. Before starting the first workshop, participants are carefully selected and motivated by the top manager. The interval of the workshop is two weeks, and all three workshops in a series of events should be finished in three months. These observations are similar to other cases.

5 Discussion

In this paper, we proposed the Innovation Support Technology (Innovation Support Technology) that combines the METI-Strategic Technology Roadmap (METI-TRM) with the new business planning method, integrating the Business Modeling Method and the Strategic Roadmapping.

1. The Business Modeling Method creates new business value and draws up a more reliable operation plan.
2. The Strategic Roadmapping represents a new technology development plan and/or new product-development plan that take into account social trends, resource conditions, and so forth. It can easily find and eliminate various discrepancies between development schedules on a time axis.
3. Moreover, we can obtain the following benefits by utilizing the METI-strategic technology roadmap (METI-TRM) that is the summarization of professionals' wisdom in various technology fields: Our business plan is ensured by technology trends described in METI-TRM and its realizability is confirmed.

Learning about technology trends and their achievement levels in different fields enables us to explore possibilities of a new product or new market developments due to technology fusions. The Innovation Support Technology can produce more persuasive business execution plans, that is, the integrated strategic roadmap, by integrating the business model and the strategic roadmap. Further merit is that the METI-Strategic Technology Roadmap (METI-TRM) provides valuable technological information for making decisions about our business model through own company's roadmapping procedure. This fact fills a gap and reinforces reliability and realizability of our business model. The integrated strategic roadmap clearly indicates the best timing for investment and/or market introduction in order to bring attractive products to the market.

In this paper, we did the business modeling and the strategic roadmapping for a visually impaired person walking-support system as the second start-up business plan, which is an example in the real industrial world. As a result, we confirmed that the Innovation Support Technology is extremely useful.

In this case, there was initially a big gap between "my vision & my will" which the person in charge could depict and that of the company's long-term management vision. We succeeded in designing the business model that expands the current sales amount to 10 times bigger business (4 billion yen/year) and, therefore, in eliminating the initial gap.

It was demonstrated that by using the Innovation Support Technology that incorporates the strategic roadmapping into the business modeling, a more powerful business execution plan could be developed, where their weaknesses and strengths are covered and enhanced with each other.

6 Conclusions

In conclusion, using the integrated strategic roadmap, quality improvement in decision-making is realized, since the new business plan design framework provides clear and more practical strategies and enables multiphase examinations on a time axis.

What problems we observed during preparation and implementation in each of our three workshop steps and what critical success factors we derived out of these problems for each of the steps are just to describe "my vision & my will". My vision & my will" was hardly derived by the workshop leader alone. Careful support and advice by the program leader are essential for the workshop leader to make a clear concept of "my vision & my will". Once "my vision & my will" was shown, business modeling and roadmapping are routine tasks rather than challenging work. These observations are similar in these other cases.

Acknowledgement. These research activities were organized in Technology Management Research-working group of the Japan Techno-Economics Society (JATES). The author would like to express their gratitude to the JATES members who contributed their ideas through discussions and the JATES management office for supporting the research activities. The author would especially like to thank Dr. G. Trauffer and Dr. H. Sakuma for valuable comments and discussions.

References

Abe, H., Ashiki, T., Suzuki, A., Jinno, F., Sakuma, H.: Integration Studies of Business Modeling and Roadmapping Methods for Innovation Support Technology (IST) and its Practical Application to Real-World-Cases. In: Proceedings of Portland International Conference on Management of Engineering and Technology, PICMET 2007 (2007)

Abe, H., Hirabayashi, Y., Horiuchi, T., Kado, M., Sakuma, H.: A New Framework of Business Modeling Method for R&D Outputs: Valuation and Communication Tool for Engineers, Managers and Investors. In: PICMET 2004 (2004)

Abe, H., Hirabayashi, Y., Ishida, F., Oku, Y., Kado, M., Sakuma, H.: Value Creation Framework of Business Modeling Method for R&D Output. In: PICMET 2005 (2005)

Abe, H., Shinokura, K., Suzuki, A., Kubo, H., Sakuma, H.: 2nd Generation Business Modeling: Smart Innovation Planning Method Managing the Link to Corporate Value Creation for R&D Outputs. In: Proceedings of Portland International Conference on Management of Engineering and Technology, PICMET 2006 (2006)

Assistive Service Business of Ikeno Tsuken Co., Ltd. (2011), Retrieved World Wide Web, http://www.ikeno.co.jp/fukushi/index.html (accessed December 22, 2011)

Bucher, P.: Roadmapping: Some additional remarks and examples. In: Global Advanced Technologies. Innovation Consortium Roadmapping Seminar (2002)

Chesbrough, H.: Open Innovation - The New Imperative for Creating and Profiting from Technology. Harvard Business School Press, Boston (2003)

Hamel, G.: Leading the Revolution. Harvard Business School Press, Boston (2000)

Heijden, K.: Scenarios: The Art of Strategic Conversation. John Wiley & Sons, Chichester (1996)

Ikeda, K., Imaeda, M.: Scenario Planning. Toyo Keizai (2002) (in Japanese)

Ishida, F., Sakuma, H., Abe, H., Fazekas, B.: Remodeling Method for Business Models of R&D Outputs. In: Proceedings of Portland International Conference on Management of Engineering and Technology, PICMET 2006 (2006)

Kameoka, A.: The IT industry and MOT. The Symposium on Future Electron Devices (2003)

MacMillan, I.C.: Discover Driven Planning. Harvard Business Review (July/August 1995)

Ministry of Economy, Trade and Industry, The Strategic Technology Roadmap (2006), http://www.meti.go.jp/policy/kenkyu_kaihatu/18fy-pj/I-plan.pdf (in Japanese) (accessed December 22, 2011)

Phaal, R., Farrukh, C.J.P., Probert, D.R.: Developing a Technology Roadmapping System. In: PICMET 2005 (2005)

Porter, M.: Competitive Strategy: Techniques for Analyzing Industries and Competitors. Free Press, New York (1980)

Porter, M.: Competitive Advantage: Creating and Sustaining Superior Performance. Free Press, New York (1985)

Porter, M.: Michael E. Porter on Competition. Harvard Business School Press, Boston (1998)

Robert, O., Smith, J.: Influence Diagrams, Belief Nets and Decision Analysis. John Wiley & Sons, Chichester (1990)

Slywotzky, A.J.: Value Migration. Harvard Business School Press, Boston (1998)

Slywotzky, A.J.: The Profit Zone. Times Business, New York (1997)

Slywotzky, A.J.: The Art of Profitability. Warner Books, New York (2002)

Tschirky, H., Jung, H.-H., Savioz, P.: Technology and Innovation Management on the Move - From Managing Technology to Managing Innovation-driven Enterprises. Orell Füssli Verlag, Zürich (2003)

Author

Hitoshi Abe moved from OKI to JATES as Secretary General Executive' Committee for the Management of Technology in 2007. He is the leader of a study group of JATES on this issue from 2002. He received Doctor of Science from Berlin Free University in 1978. From 1978 to 1981, he continued research on chemical physics at Fritz Haber Institute of Max Planck Society in Berlin. In 1982, he joined Oki Electric Industry Co., Ltd. He was engaged in research on new function devices and in R&D management. From 1996 to 2000, his mission moved to corporate management and technology management. From 2000 to 2007 he was CTO (Chief Technology Officer) Associate in OKI. Fields of his current interest are research and development on the practical management tools for innovation and technology.

Part 3: Implementing Technology Roadmapping

To exploit the benefits of technology roadmapping to its full extent it is important to maintain and support the process, which requires consideration of institutional aspects, allocating staff and other resources in an efficient manner. Procedures and other related measures in decision-making need to be carried out efficiently so that roadmapping may finally become an integral part for the management of technology and innovation, and not just a standalone technical tool used merely for a distinct roadmapping project.

In this part of the book promising ways of how technology roadmapping can be implemented and institutionalized are described. This includes software tools, taking into account the dynamics of key players, different levels of involvement, illustrating a variety of roles, activities and responsibilities for technology roadmapping so that it aligns with and supports strategy and innovation in the firm.

Four implementation processes are presented in this chapter:

- *Gerdsri* focuses on the implementation of technology roadmapping as a complex process for the organization, exploring roles, actions and success factors. The example of a leading manufacturer in Southeast Asia illustrates the motivation and approach for implementing technology roadmapping. The starting point of this contribution is the classification of the implementation process into three stages: initiation, development and integration.
- *Farrokhzad, Kern and de Vries* provide an overview of the portfolio-based roadmapping process at Siemens AG, and its incorporation into the organization. The starting point of their contribution is the Innovation Business Plan which helps the company to improve overall evaluation and selection abilities, enhancing innovation performance.
- *Beeton, Phaal and Probert* present a standardized roadmapping process that was developed in an industrial project to produce the International Roadmap for Consumer Packaging. The view is on an exploratory roadmap for sector-level foresight. The approach shows how several companies or institutions can collaborate to explore future markets, products, technologies and resource trends. The starting point is to establish a steering committee which is responsible for the design and implementation of the process.

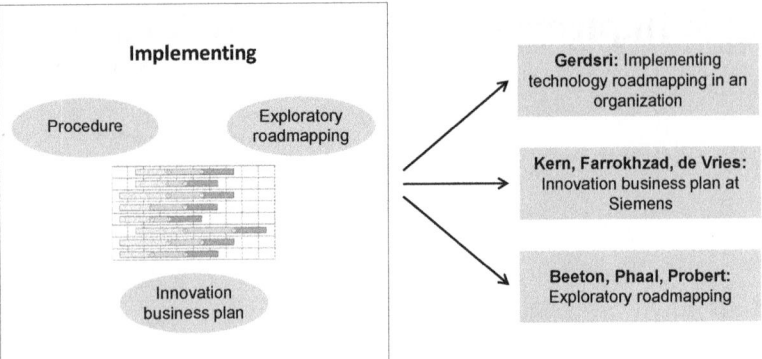

Fig. 1 Implementing technology roadmapping

Implementing Technology Roadmapping in an Organization

Nathasit Gerdsri

With the completion of technology roadmapping implementation, an organization can be assured that its required technologies and infrastructures will be ready when needed. Implementing technology roadmapping as a part of the ongoing strategic/business planning process is challenging because the consequences of technology roadmapping implementation can lead to some changes in the business process, organizational structure, or even working culture. Therefore, an organization needs to understand how the changing roles and responsibilities of key players involved in the technology roadmapping process match with the dynamics of technology roadmapping implementation in each stage. This chapter illustrates the dynamics of technology roadmapping implementation and the importance of applying a change management approach to address an individual' needs and challenges in adopting the technology roadmapping process. A case example is also presented to demonstrate how one of the leading building product manufacturers in the ASEAN region manages through the dynamics of technology roadmapping implementation.

1 Introduction

Some organizations implement technology roadmapping to guide their strategic vision. Other organizations implement technology roadmapping more extensively to lead their operations. For the latter case, organizations usually integrate technology roadmapping as a part of their strategic planning activities to link market opportunities and potential products/services with the development of proper technologies to support the future needs. In this case, the roadmapping results also link to a resource allocation plan.

In either case, organizations have to keep their roadmap up-to-date. To do so, an organization must plan the activities to periodically verify its roadmaps and then, if necessary, adjust them in a timely manner. Thus, the process of technology roadmapping should be integrated as a part of an organization's ongoing business process so that a roadmap can be kept alive.

Nathasit Gerdsri
College of Management, Mahidol University,
69 Vipawadee Rangsit Rd., Phayathai Bangkok, Thailand
e-mail: cmnathasit@mahidol.ac.th

There are several studies addressing some obstacles and challenges in implementing technology roadmapping in an organization. For example, an organization may face challenges about the issues of *start-up a technology roadmapping process* (Farrukh, et al., 2001; Groenveld, 1997, 2007; Phaal, et al., 2003a, 2003b), *top-management commitment* (Groenveld, 1997, 2007; Kappel, 2001; Kostoff and Schaller, 2001; McMillan, 2003), *selection of the right players* to involve in the technology roadmapping implementation (Brown and O'Hare, 2001; Gerdsri and Vatananan, 2007; Gerdsri, et al., 2009), and *choosing and customizing a correct technology roadmapping approach* (Fleury, et al., 2006; Gerdsri, et al., 2009; Hoffman and Daim, 2006; Holmes and Ferrill, 2005; Kostoff and Schaller, 2001; Lee and Park, 2005; Phaal, et al., 2001; Probert, et al., 2003). Also, due to limited data availability of new and emerging technologies or market forces, an organization may face challenges in *predicting future events* (Gerdsri, 2005, 2007; Strauss and Radnor, 2004; Vojak and Chambers, 2004). Lastly, an organization could find it challenging to *facilitate workshops to generate and share the knowledge needed for roadmapping* (Phaal, et al., 2007; Gerdsri, et al., 2009).

For an organization that aims to implement technology roadmapping as a part of its strategic planning activities, it is vital to integrate the technology roadmapping process into ongoing business operations so that the survival of technology roadmapping implementation can be successfully assured (Farrukh, et al., 2001; Gerdsri and Vatananan, 2007; Gerdsri, et al., 2009; Groenveld, 1997, 2007; Phaal, et al., 2001, 2004, 2005; Rinne and Gerdsri, 2003; Strauss and Radnor, 2004). Going through the technology roadmapping integration, the process is complex and may lead to some major changes in an organizational structure and culture (Cosner, et al., 2007; Gerdsri, et al., 2008, 2010; McMillan, 2003). Therefore, it is important to apply appropriate change management techniques (Cosner, et al., 2007; Gerdsri, et al., 2008, 2010) and proper training to prepare all key players to be ready for the technology roadmapping implementation (Gerdsri, et al., 2009, 2010; McMillan, 2003).

2 Stages of Technology Roadmapping Implementation

Gerdsri et al. (2009) proposed to classify the general approach for technology roadmapping implementation in an organization into three stages: initiation, development, and integration (Gerdsri and Dansamasatid, 2008; Gerdsri and Vatananan, 2007).

Stage 1 **Initiation stage** aims to get an organization ready before beginning to implement technology roadmapping process.

Stage 2 **Development stage** aims to develop a desired roadmap by engaging right people, gather the necessary information, and conduct a step-by-step analysis.

Stage 3 **Integration stage** aims to integrate technology roadmapping process into an ongoing business planning activities so that a roadmap can be constantly reviewed and updated in a timely manner.

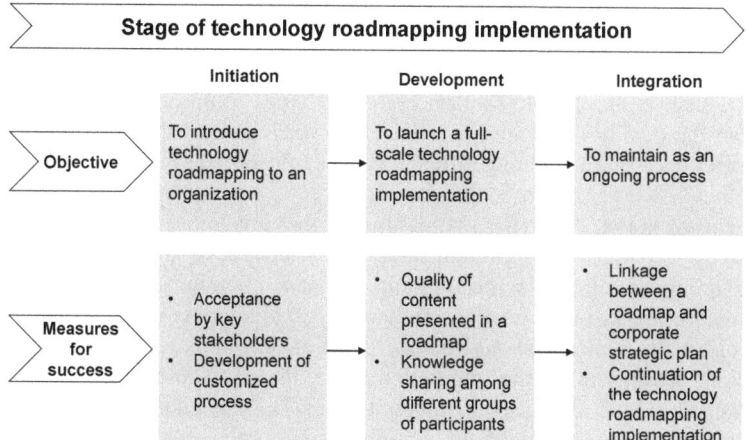

Fig. 1 Objective and measures for success in different stages of technology roadmapping implementation

For organizations that intend to develop a roadmap as a one-time effort for guiding their strategic vision, the technology roadmapping implementation effort can be stopped after the development stage is completed. However, some organizations, which want to assimilate the technology roadmapping process into their ongoing business operations, need to continue through the integration stage. The detailed explanation of main activities conducted in each stage and the key measures for success are described as follows and summarized in Figure 1.

Initiation stage: It is important for any organization to start off with the right approach. The purpose of this stage is to provide the opportunity for gathering and disseminating necessary information to use in the later stages. In this stage, core teams are formed. The individuals as well as groups prepare themselves by understanding the basic knowledge, requirements, and approach of technology roadmapping. The ground rules for team participation need to be set as well.

After the official kick-off for the technology roadmapping initiative, basic information addressing the concept of technology roadmapping is distributed to key stakeholders, in order to convince them and get them to buy into the initiative. With increasing numbers of supporters and buy-ins, the first-cut technology roadmapping workshop can be organized. Throughout the initiation stage, the core teams should discuss about the customization of generic technology roadmapping concept to make it fit into the strategic planning process and the organization's working culture.

The success of activities in this stage can be measured through the acceptance of technology roadmapping concept among key stakeholders and the customization of technology roadmapping process to meet organizational needs.

Development stage: The main emphasis of this stage is on data collection and analysis. A series of technology roadmapping workshops is conducted to analyze collected data and graphically present the results in a roadmap form. The workshop participants are the members of the technology roadmapping operation team organized within each strategic business unit (SBU). The collection of data can be done both internally and externally. The benefits gained from the workshops are not only to analyze data, but also, to share, transfer and create knowledge (Nonaka, 1991; Phaal, et al., 2005). The success of activities in this stage can be measured through the quality of content presented in a roadmap and the level of knowledge and experience sharing among different groups of participants.

Integration stage: After the completion of the development stage, the focus of technology roadmapping implementation is moved to the integration of the roadmapping process into the ongoing business operations of the organization. This integration is vital, since the technology roadmapping initiative is not a one-time effort but rather it should be exercised as an ongoing process (Kostoff and Schaller, 2001). During the integration stage, the main roles and responsibilities are transferred to the idea champion team. The aim and desired result is the complete fusions of the technology roadmapping process into the organization, so that the roadmapping process becomes a part of strategic business planning. With the successful integration, a roadmap will be maintained and updated as part of normal business operations.

The success of activities in this stage can be measured through the strength of the linkage between technology roadmaps and a corporate strategic plan as well as the continuation of technology roadmapping implementation.

3 Key Players Involving in Technology Roadmapping Implementation

In general, the contributions from individuals and teams are necessary to assure the successful implementation of any initiative in an organization. This is also applicable for the technology roadmapping implementation. Key players involved in the technology roadmapping implementation come from different levels and sub-groups of the organization. The most important and influential players are idea champions, champion team, technology roadmapping operation team and support team. The engagement from an external consulting team may be necessary, especially in an organization that implements technology roadmapping for the first time. The involvement of the external consulting team may vary from one organization to another and the level of its engagement should be carefully assessed to assure that technology roadmaps can be developed and the technology roadmapping process can be integrated into the ongoing business operations. The following section describes the characteristics of each player and their interaction, which can be conceptually illustrated as shown in Figure 2.

Idea champion: The idea champion is the one who provides the energy to move the subjects to gain acceptance for the change (Daft and Bradshaw, 1980). The emergence of the idea champion is an indispensable ingredient in the process of innovation and strategic change (McCall and Kaplan, 1985). This individual sees not only the needs and benefits for innovation, but also provides transformational leadership throughout the implementation process (Howell and Higgins, 1990).

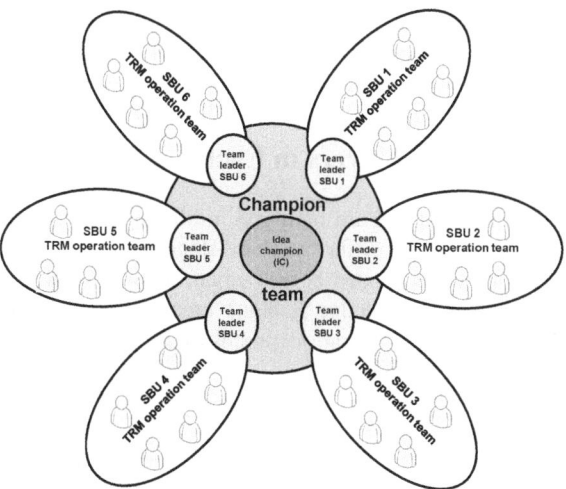

Fig. 2 Interaction among key players (SBU: strategic business unit)

This makes the idea champion important as an impetus to change and to overcome constraints. The champion's main role is to guide the members of the organization through the change process and mitigate constraints along the way.

Champion team: The champion team is a group of idea champions in which each individual represents a different business unit. As compared to the characteristics of individual idea champions, the champion team makes use of its team dynamic, commitment, diversity, and flexibility in addition to their individual traits (Kotter, 1996). The champion team is a driving force of the technology roadmapping initiative and is critical to its success. Each member of the team will lead a group of participants from his/her business unit through the technology roadmapping development process. In addition to the roadmap development, the champion team is also responsible for finding a proper way to integrate the roadmapping process into the ongoing business processes of the organization.

Technology roadmapping operation team: The technology roadmapping operation team is a working group assigned and handpicked by the idea champion of each business unit, to participate in the development of a technology roadmap. An idea champion who is also a member of the champion team leads the group. The members of the technology roadmapping operation team are recruited from major divisions (e.g. strategic planning, marketing, engineering, product development) of each business unit. Each individual has strong knowledge and experience in his/her functional area. With the combined knowledge from the team, the future trends and needs for the business unit can be determined and used as strategic inputs for the technology roadmapping development.

Technology roadmapping support team: The technology roadmapping support team is formed and initiated by the idea champion as the administrative body of the technology roadmapping implementation. Even so, the role of this team will increase over the course of the implementation. Its main function is to capture, store and distribute resources and information. The support team will be the resource center of the technology roadmapping initiative.

4 Involvement of Key Players in Each Stage

All aforementioned key players are involved in all stages of implementation but the levels of their involvement differ from one stage to another. In each stage, the types of engagement can be scoped at individual, team as well as organizational level. Through all stages, knowledge is being created, shared and transferred as the driving part of the implementation process (Gerdsri and Vatananan, 2007; Gerdsri, et al., 2009; Senge, 1990).

4.1 Initiation Stage

During the initiation stage the focal roles of individuals and teams are to learn and to communicate. The main responsibility is to understand the complete process of technology roadmapping and the nature of the three implementation stages. The following section will discuss the roles and responsibilities of the key players in this stage.

The idea champion: Because the technology roadmapping process represents a considerable change to the organization, a single idea champion might not be able to cope with the amount of responsibilities. This is why the idea champion needs to communicate his/her knowledge and understanding of the initiative to other key stakeholders, in order to form a team of idea champions (also referred by Kotter as a guiding coalition) (Kotter, 1996). In this stage the champion's main responsibility is networking to bring the right people together (Phaal, et al., 2003a). To do this, his/her individual mastery on technology roadmapping implementation is essential in order to communicate and inform or even educate key stakeholders.

The champion team: The group of individual idea champions, assembled from several strategic business units, forms the champion team. The team members are well respected experts who will lead their strategic business units through the implementation process. This high performance team will be the driving force through the three stages of the technology roadmapping implementation. The formation of the idea champion team will ensure the effective diffusion of how important the technology roadmapping initiative is for the organization (Kotter, 1996). But to be effective communicators, the members of the team need to have the same level of understanding and knowledge about the initiative. Through individual learning and first-cut workshops, all members of the champion team will raise their level of understanding about technology roadmapping. The common and equal understanding among the team members creates a dynamic, which is needed for the collective learning and facilitation process

(Hosley, et al., 1994; Senge, 1990). In addition to their function as transformational leader and facilitator of the technology roadmapping development, the champion team will serve as a communication link between Top management and the technology roadmapping operations team (Nonaka, 1991).

Technology roadmapping operation team: For the purpose of a roadmap development, the individual member of the champion team, who is a representative from each strategic business unit, is responsible for setting up his/her own technology roadmapping Operation Team. Members of the team must be carefully chosen to represent an effective blend of expertise across several functional departments within the strategic business unit (e.g. Marketing, Engineering, Product Development, R&D, Finance, etc.) (Probert & Shehabuddeen, 1999). At this stage, the activities of the technology roadmapping operation team are mainly to capture the basic knowledge and get themselves ready for participating in the development stage.

Technology roadmapping support team: The technology roadmapping support team begins to form with the objective to operate as a resource center. The team's responsibility is to provide basic information related to their corporate business and technology roadmapping implementation collected from both internal and external sources. The team needs to create open-communication channels for all strategic business unit's technology roadmapping operation teams to access that information.

4.2 Development Stage

At this stage, the actual roadmap is being developed. The major activities are to extract, distribute and share knowledge through a series of workshops (Phaal, et al., 2003a; Phaal, et al., 2005). The following sections will discuss the roles and responsibilities of the key players in this stage.

The idea champion: At the beginning of the development stage, the idea champion assumes the role of a leading facilitator, responsible for the coordination of relevant knowledge sources and preparation of individuals for the roadmap development. The idea champion needs to create a relaxed and productive atmosphere for participants in order to smooth the progress of knowledge creation and transfer which will continuously incur during the technology roadmapping development process. The idea champion also conducts feedback sessions with internal and external participants to make appropriate adjustments. Once the champion team is ready to carry the work, the idea champion will transfer the ownership to the team and then focus his/her roles on assisting the champion team as conflicts and problems arise.

The champion team: After the idea champion provides general guidance to the champion team, each member of the champion team starts to facilitate the technology roadmapping development process for his/her own strategic business unit. Being a part of the champion team, each member is responsible to share feedbacks and lessons learned from the workshops with all members so that the team can strengthen their common knowledge and experiences. In addition, the individual member of the champion team in each strategic business unit must play a role as a project manager in controlling the progress, motivating participants and negotiating for proper resources.

Technology roadmapping operation team: In this stage, the technology roadmapping operation teams from all strategic business units carry the main load in developing their unit's roadmap. The team members gather and analyze relevant information needed through the roadmap development process. The gained insights are captured and discussed among all team members so that the team can draw their conclusions for the final decisions. As a result, these activities can be considered as a mean to convert tacit into explicit knowledge as mentioned by Nonaka (1991). During this stage, it is important for the technology roadmapping operation team to engage themselves in open dialogs to enhance the team's learning process (Hosley, et al., 1994; Senge, 1990).

Technology roadmapping support team: Being a resource center, the technology roadmapping support team must assemble all necessary information and maintain all communication channels open, so that the information can be shared and exchanged.

4.3 Integration Stage

The development of a technology roadmap alone is not enough to sustain it. The roadmap implementation needs to be seen as an ongoing process with continuous review and update of drivers, technologies and the map itself (Kostoff and Schaller, 2001). Therefore the integration of roadmapping processes into the current business operation in any organization is the key to sustainability. The following section will discuss the roles and responsibilities of the key players in this stage.

The idea champion: The final role of the idea champion is to initiate the integration process of the technology roadmap into an ongoing business process. The sole responsibility of the idea champion is to oversee the integration and continue to provide assistance and support to the champion team. Finally, with completion of the technology roadmapping integration, the role of the idea champion changes into a technology roadmapping Sponsor, who is no longer involved in the process, but will remain in an advisory capacity.

The champion team: With the decreasing role of the idea champion, the champion team will pick up the pace of integrating the roadmap into the organization's ongoing business processes. It might be necessary to redesign or remove redundant processes along the integration. With the successful integration, the challenge of the champion team is to keep the technology roadmapping process alive. Therefore, in addition to the roles and responsibilities as described above, the individual champion must be in charge of preserving the roadmap and ensuring its continuation.

Technology roadmapping operation team: At the beginning of this stage, the team will assist the champion team in verifying and validating their roadmap. As the roles and responsibilities of technology roadmapping operation team gradually decline, they still help the champion team to create linkages between technology roadmapping process and the current business processes of the organization. With successful completion of the integration stage, the technology roadmapping operation team is embedded into the organization's ongoing business processes. Finally, the team's roles and responsibilities change to update and adjust their technology roadmap.

Technology roadmapping support team: The support team is responsible for preparing and collecting documents produced by the technology roadmapping development process. Apart from these tasks, the team also assists the champion team in compiling all strategic business unit roadmaps into a corporate master roadmap. Lastly, the team has to diffuse the knowledge and lessons learned from the process throughout the organization. The idea of knowledge diffusion in this stage can be referred to Nonaka's concept on knowledge creation where tacit knowledge must be transformed into explicit and is then distributed throughout the organization (Nonaka, 1991).

5 The Dynamics of Technology Roadmapping Implementation

The dynamics of technology roadmapping implementation can also be addressed by the change on the degree of involvement of each key player. Gerdsri et al. (2008) observed how the level of involvement of each key player gets change along the three stages of technology roadmapping implementation. The patterns of the change on the degree of involvement of each key player can be conceptually illustrated as shown in Figure 3.

During the initiation stage, the level of involvement of the idea champion is high as the idea champion is the one who provides the momentum for the implementation, by forming key teams and training them to be ready for the technology roadmapping implementation. While other key players begin to learn and understand the technology roadmapping implementation process, the idea champion is communicating with key stakeholders to promote the concept of technology roadmapping to secure support and vital resources.

In the development stage, the idea champion hands over most of the responsibilities to the champion team and assumes a supportive and advisory role. His/her role is to monitor the technology roadmapping development and, if necessary, adjust the process or material. As the technology roadmapping development commences the involvement of the operation team peaks. This is the stage where knowledge is being created and transferred, to formulate the roadmap.

Once the roadmap has been developed and the integration stage has begun, the idea champion assumes the role of a technology roadmapping advisor. Now the focus lies on the champion team and technology roadmapping support team to compile all the business unit roadmaps into a master roadmap and to integrate the

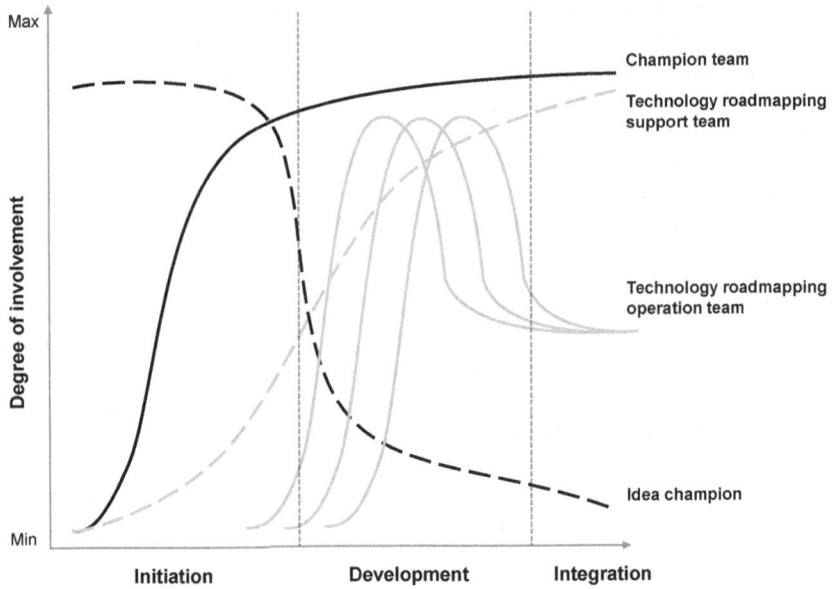

Fig. 3 Conceptual illustration of the changes on the degree of involvement for each key player throughout different stages of the technology roadmapping implementation

technology roadmapping process into the ongoing business processes of the organization. In this stage, the level of involvement of the champion team and technology roadmapping support team reach the highest level. While the level of involvement of technology roadmapping operation team slightly decreases as their roles reduces from developing the roadmap to assisting the integration effort. With successful integration, each idea champion of the champion team will assume ownership of the technology roadmapping process in their corresponding business unit and lead the technology roadmapping operation team in maintaining and updating the roadmap as part of their daily operations.

6 Applying Change Management to Guide Technology Roadmapping Implementation

As mentioned earlier, technology roadmapping is not just a process; people involved in the process are crucial to the success as well. Applying change management can assist an organization to set up a proper technology roadmapping implementation plan and guide key players to cope with the new technology roadmapping process. Two well-known concepts of change management, Prosci's ADKAR model and Kotter's eight stages of change, can be taken into consideration for the development of an activity guideline for technology roadmapping implementation.

6.1 Prosci's ADKAR Model

The ADKAR model focuses its attention on the five elements to prepare individuals for the change. Each element of change is described below (Hiatt, 2006):

Element 1 **Awareness of the need for change:** To be able to change, the organization and its members need to know and understand the rationales for changes. Awareness represents a person's understanding of the nature of the change including why the change is being made and the risk of not changing.

Element 2 **Desire to make the change happen:** To motivate people to change, the organization needs to create positive or negative consequences influencing individual's desire for engaging in a change. Desire represents the willingness to support and engage in a change. Desire is ultimately about a personal choice influenced by the nature of the change as well as an individual's personal situation and intrinsic motivation.

Element 3 **Knowledge about how to change:** Motivation to change alone is not enough to initiate a change. Individuals need to understand what the proper behavior looks like. They need examples and guidance so that they can obtain the knowledge of what the correct behavior or procedure is.

Element 4 **Ability to implement new skills and behaviors:** Once the required knowledge is obtained, basic practices need to be provided to attain the abilities and skills necessary for engaging in a change. Ability can only be achieved when a person or group has the demonstrated capability to implement the change at the required performance levels.

Element 5 **Reinforcement to retain the change:** To anchor the new behavior in the corporate culture, individuals need some reinforcement to keep the good behavior going. Reinforcement consists of both internal and external factors. External reinforcements could include recognition, rewards and celebrations that are tied to the realization of the change. Internal reinforcements could be a person's internal satisfaction with his or her achievement or other benefits derived from the change on a personal level.

6.2 Kotter's Eight Stages of Change

The eight-step process is a direct response to the top eight most common errors that organizations and management makes when confronted with change (Kotter, 1995, 1996). A brief definition of each stage is provided below (Cohen, 2005):

Stage 1 **Establishing a sense of urgency (increase urgency):** The prime objective of this stage is to raise awareness of the need and importance for changes. If it is done right, the creation of urgency atmosphere will reduce complacency and gain needed cooperation. It will generate interest and motivate people to take action.

Stage 2	**Creating the guiding coalition (building guiding teams):** A guiding coalition will facilitate the decision-making process, access more information and be able to act more quickly. A synergistic effect among members can be enhanced by providing some assistance to a team in areas like communication and knowledge sharing.
Stage 3	**Developing a vision and strategy (get the vision right):** To be successful, the change effort needs a direction representing a compelling and motivating picture of the future. A vision helps to coordinate actions and to identify behaviors that should be encouraged or eliminated.
Stage 4	**Communicating the change vision (communicate for buy-in):** Communication of the vision is essential to develop the understanding about the necessity of the change effort and to convince people to buy-in. This helps to gain access to needed recourses and captures the commitment from workforce. By sharing the desired future, it helps motivate and coordinate all participating members to go through the transformation.
Stage 5	**Empowering employees for broad-based action (enable action):** To carry out the change, each individual in an organization need to be enabled to take broad base action.
Stage 6	**Generate short-term wins:** These timely, visible, and meaningful achievements are critical to build the credibility needed to sustain the change effort over time. They also provide a visible proof to stakeholders that the effort pays off. As a result, motivation, moral and commitment of key players can be maintained.
Stage 7	**Consolidating gains & producing more change (don't let up):** The momentum created by the short-term wins can be used to move the change effort forward, and enables key players to take on bigger and deeper changes. However, it is essential for leaders to continue conveying their commitment to employees and management as well as keep up urgency and not to declare premature victory.
Stage 8	**Anchoring new approaches in the culture (make it stick):** In this stage the new behaviors are woven into the organizational culture. In order to achieve a sustainable integration of the change, leaders need to adopt the new behaviors themselves, as well as reward and recognize their subordinators in adopting the new behaviors.

6.3 Applying ADKAR Model and Eight-Step Model to Develop an Activity Guideline for Technology Roadmapping Implementation

The combination of ADKAR and Kotter's 8-stages model of change management will aid technology roadmapping practitioners to understand the basic elements needed in preparing individuals for changes and the actions needed in managing the technology roadmapping process through all three stages. Table 1 presents the

Table 1 An activity guideline for technology roadmapping implementation

Change management concepts		Action plan supporting the three-stage technology roadmapping implementation process	Stage of technology roadmapping implementation	Key success factors (KSFs)
Prosci's elements of change	Kotter's 8-stage of change			
Awareness Desire Knowledge	Step 1: Create a sense of urgency	Understand the value of applying technology roadmapping in the organization Build awareness of why technology roadmapping implementation is needed	Initiation	Acceptance of the initiative by key stakeholders Development of a customized technology roadmapping process
	Step 2: Form a guiding coalition	Discuss the details of technology roadmapping concept Raise urgency of why technology roadmapping implementation is immediately necessary to all participating members		
	Step 3: Develop vision & strategy	Develop a vision, objective, and scope of technology roadmapping implementation for the organization Set the plan to roll-out technology roadmapping implementation		
	Step 4: Communicate vision & strategy	Gain acceptance and sponsorship from top-management Communicate the vision for the buy-in and support from key players Form a working group responsible for activities related to technology roadmapping implementation Provide the fundamental concept of technology roadmapping to all participants Prepare all participants to be ready to implement the technology roadmapping process. Training sessions may be provided. Customize the generic technology roadmapping process to fit with the organizational setting		
Ability	Step 5: Empower people	Plan and organize a series of workshop sessions to develop a roadmap Allocate responsibilities to each individual in the group as well as set up ground rules for the group participation	Development	Content quality presented in the roadmap Knowledge sharing among different groups of participants
	Step 6: Create short-term wins	Maintain the momentum and energy from all participants throughout the technology roadmapping development process Remove barriers blocking participants from carrying out their technology roadmapping activities Conduct debriefing and review sessions		
Reinforcement	Step 7: Consolidate gains	Consolidate roadmaps into one master roadmap (if needed) Establish the procedures to review and revise a roadmap so that a roadmap can be kept alive	Integration	Linkage between roadmap and corporate strategic plan Continuation of technology roadmapping
	Step 8: Anchor new approaches	Integrate technology roadmapping process into organization's existing processes Transfer ownership of the process to the proper group of people		

examples of activities needed for facilitating and supporting technology roadmapping implementation process. These activities are properly determined corresponding both ADKAR's and Kotter's model.

As described earlier, the purpose of the initiation stage is to define the scope of the technology roadmapping initiative, select a team of key players, and prepare them to be ready for technology roadmapping implementation. The three elements of ADKAR model; Awareness, Desire, and Knowledge, are concentrated along with the Kotter's first four steps to create a sense of urgency, form a guiding coalition, develop a change vision, and communicate the change vision throughout the organization.

In the development stage, the activities emphasize the interaction of key players and their synergistic effort in analysing data and information. The interaction among individual team members can lead to the creation of new knowledge supporting the development of a roadmap. The "Ability" element of ADKAR model is concentrated along with the Step 5 and 6 of Kotter's model emphasizing theempowerment of people involved in the process and the plan for achieving short-term wins.

In the integration stage, the main attention lies on how to seamlessly integrate the new technology roadmapping process into day-to-day operations at the organization. The last element of ADKAR model, Reinforcement, is considered in conjunction with the Step 7 and 8 of Kotter's model. The emphasis is to invigorate the new processes and behaviours in the organization to anchor the technology roadmapping process as a part of normal business practice and to keep the roadmapping process alive.

It is important to remember that technology roadmapping need to be customized for each organization to fit with its organizational context. The action plan presented in Table 1 is general and can be applied for most situations.

7 Implementing Technology Roadmapping: A Case Example

In the following chapter the technology roadmapping implementation is illustrated. The company presented in this case example is one of the leading manufacturers in the ASEAN region

7.1 Background

The company has multiple strategic business units (SBUs). Its products have been widely recognized for the high quality by customers, both in local and international markets. Several market research activities have been initiated with special emphasis on fulfilling customers' requirements and satisfactions.

Recently, the company announced its new strategic vision to focus on "creating a product solution for all" instead of providing separated components to the market. To complete this new vision, the company decided to reform itself from being a product manufacturer to become a total solution provider. Therefore, the development plan for each product must be integrated into one system, in order to provide better product solutions for their customers in the future.

7.2 The Company's Objective and Motivation in Applying Technology Roadmapping Approach

Currently, the company carries out two types of business planning. One is a high-level strategic plan known as a medium-term plan (MTP). The other is an annual action plan. Medium-term plan addresses the company's strategic issues regarding its future direction for the next five years. The content of medium-term plan covers the plan for the development of new products, new markets, as well as new business operations. On the other hand, the content of annual action plans covers

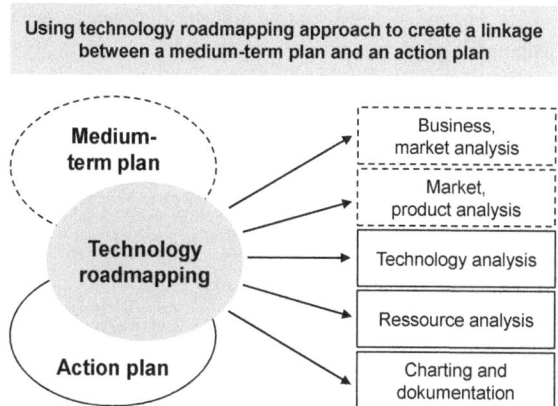

Fig. 4 The company's objective and motivation in applying in technology roadmapping

activities with estimated budget and ownership assignment. The completion of activities in annual action plans would achieve the strategic objectives as indicated in a medium-term plan for each upcoming fiscal year.

The two strategic plans have been effectively used to manage the development of the company's technological and manufacturing capabilities required for new product development. However, with the new vision of becoming an industry leader in providing a total housing system solution, several strategic business unit s must collaboratively work to assure that their products would be seamlessly integrated into a system and flawlessly function together. As a result, the company has to spend longer time in R&D before being ready to launch a new product system solution than the duration the company is used to spend when each strategic business unit develops and launches its own products, independently.

The company is considering the potentials of applying technology roadmapping in conjunction with its medium-term plan and annual action plan (as shown in Figure 4.) so that the company can emphasize identifying potential products, system, and services, map them onto technology alternatives, and develop resource allocation plans. By applying technology roadmapping approach, the company will be able to ensure that the required technologies and infrastructures will be ready when needed. Moreover, the activities conducted by various strategic business unit s in delivering their products to support a new product system could be synchronized so that the solution can be launched in a timely manner.

7.3 The Company's Approach in Implementing Technology Roadmapping

The technology roadmap development project was carried out after a three-month preparation. The company's ultimate objective was to have a corporate master roadmap representing the future direction of all major strategic business units. Table 2 describe how the subject company organized and exercised its activities to support technology roadmapping implementation along the three stages by following the activity guideline as presented in Table 1.

Table 2 Applying a change management approach to organize activities supporting technology roadmapping implementation: A case example

	Action plan to support the technology roadmapping implementation process	A case example presenting how the subject company organized the activities to support its technology roadmapping implementation
Initiation	Understand the value of applying technology roadmapping in the organization	The idea champion was identified. He was a person who saw the value of technology roadmapping and tried to bring it into the organization. He took the initiative to acquire the basic knowledge about technology roadmapping. He communicated with the external people who have experiences in technology roadmapping to discuss about the possibility of implementing technology roadmapping at the company. After all details of technology roadmapping were clarified, the company decided to develop a product-technology roadmap for each strategic business unit and then aimed to compile them at the end to represent a corporate master roadmap. The external technology roadmapping research/consulting team was officially brought in. The research/consulting team began to work with the internal team to setup the plan for facilitating the technology roadmapping implementation process. The joint team also discussed the link between the proposed roadmapping activities and other existing strategic/business planning in the company.

To effectively roll out the activities and build up a quick buy-in from key stakeholders, the joint team decided to execute technology roadmapping implementation in multiple phases instead of rolling it out to all strategic business units at the same time. Since it was the first time that technology roadmapping was introduced to the company, setting the pilot execution allowed the joint team to fully focus on the needs of the pioneering strategic business units. As a result, the resistance from key stakeholders in adopting the new process was minimized and a quick win could be declared. The success in the pioneering strategic business units helped to reinforce the buy-in from the remaining strategic business units.

As part of building a vision and strategy for the technology roadmapping initiative, the idea champion and research/consulting team worked together to communicate with key stakeholders to rally for their buy-in and support. Due to the novelty of the technology roadmapping initiative at the company, the idea champion had to communicate with top management regarding the balance of the expectations between a learning process and the quality of the roadmap content. A kick-off meeting for technology roadmapping implementation was organized. The President of the company was invited to inform his staffs about the needs and expected benefits from the technology roadmapping implementation. |
	Build awareness of why technology roadmapping implementation is needed	
	Discuss the detailed concept and the roll-out plan of technology roadmapping implementation	
	Raise urgency of why technology roadmapping implementation is immediately necessary to all participating members	
	Develop a vision, objective, and scope of technology roadmapping implementation for the organization	
	Gain acceptance and sponsorship from top-management	
	Communicate the vision for the buy-in and support from key players	
	Form a working group responsible for activities related to technology roadmapping implementation	
	Provide the fundamental concept of technology roadmapping to all participants	
	Prepare all participants to be ready to implement the technology roadmapping process	
	Customize the generic technology roadmapping process to fit with the organizational setting	
Development	Plan and organize a series of workshop sessions to develop a roadmap	After identifying pioneering strategic business units, the technology roadmapping operation teams were formed for each strategic business units. The team leader was appointed to be a project manager of each strategic business unit team. The operation team of 6-8 members was assembled from key staffs involved in strategic planning, marketing, product development, engineering, and R&D activities in each strategic business unit.

The research/consulting team customized the technology roadmapping process to match with the types of information available. In case that the needed information was not available, the research/consulting team gave advices to the technology roadmapping operation teams to temporarily substitute that information with the team's judgment during the workshop sessions. However, the substituted information needs to be replaced later on with complete information. This approach helps keeping the momentum of the teams alive to continue through the technology roadmapping process without any interruption. A step-by-step workbook with examples was also distributed as part of the workshops. The facilitator and members of the research team helped creating a dynamic atmosphere during the workshop sessions, to allow all members of the working team to challenge each other on related issues. |
	Allocate responsibilities to each individual in the group as well as set up ground rules for the group participation	
	Maintain the momentum and energy from all participants throughout the technology roadmapping development process	
	Remove barriers blocking each participant from carrying out the technology roadmapping activities	
	Conduct debriefing and review sessions	
Integration	Consolidate roadmaps into one master roadmap (if needed)	After completing the development stage, the ownership of the technology roadmapping process was transferred from the initial idea champion to two key stakeholders in the company. One is a business planning manager, who is a process owner of business planning activities. The other one is a technology manager who manages a portfolio of technologies and leads a group of technological experts in the company. The transfer of ownership assured that the technology roadmapping process would be adopted as a part of the company's ongoing process by seamlessly integrating it into the existing business planning process. Some redundant activities between roadmapping and business strategic planning were removed so that the whole analysis process was smoothly integrated. The periodical review and update of roadmaps were scheduled by linking the timing with the development of strategic plan organized in Q3 and the development of action plan organized in Q4.
	Establish the procedures to review and revise a roadmap so that a roadmap can be kept alive	
	Integrate technology roadmapping process into organization's existing processes	
	Transfer ownership of the process to the proper group of people	

8 Conclusion

The implementation of technology roadmapping is a complex process for the organization. The classification of the implementation process into three stages (initiation, development and integration) helps the key players to understand the unique requirements and the level of involvements in each stage. To exploit the dynamics of technology roadmapping implementation, it is essential to get key players involved at different levels across multifunctional departments. Roles and Responsibilities of each player vary throughout the process. Therefore, to understand what their involvement should be focused on and how their involvement will evolve over time is critical for the key players. With the clear understanding of the dynamic linkage and relationship among individuals and groups, technology roadmapping implementation can be strengthened and knowledge can be more effectively shared and transferred. As a result, this will leads to a higher chance of a successful technology roadmapping implementation.

The proposed guideline (Table 1) could be used as a checklist for an individual or team who is responsible for deploying the roadmapping process. The guideline was developed by applying the approach of change management and addressing them through the three stages of technology roadmapping implementation. By following the proposed guideline, technology roadmapping practitioners can prepare participating members to be ready to cope with the new processes and procedures. The step-by-step action plan can lead key players and stakeholders through the changes required for technology roadmapping implementation.

References

Brown, R., O'Hare, S.: The use of technology roadmapping as an enabler of knowledge management. Paper Presented at the IEE Seminar on Managing Knowledge for Competitive Advantage (2001)

Cohen, D.S.: Heart of Change Field Guide: Tools and Tactics for Leading Change in Your Organization. Harvard Business School Press, Cambridge (2005)

Cosner, R.R., Hynds, E.J., Fusfeld, A.R., Loweth, C.V., Scouten, C., Albright, R.E.: Integrating Roadmapping Into Technical Planning. Research Technology Management 50(6), 31 (2007)

Daft, R.L., Bradshaw, P.J.: The Process of Horizontal Differentiation: Two Models. Administrative Science Quarterly 25(3), 441 (1980)

Farrukh, C.J.P., Phaal, R., Probert, D.R.: Industrial practice in technology planning - implications for a useful tool catalogue for technology management. Paper Presented at the Portland International Center for Management of Engineering and Technology, PICMET (2001)

Fleury, A.L., Hunt, F., Spinola, M., Probert, D.: Customizing the Technology Roadmapping Technique for Software Companies. Paper Presented at the Portland International Center for Management of Engineering and Technology (PICMET), Istanbul, Turkey (2006)

Gerdsri, N.: An Analytical Approach to Building a Technology Development Envelope (TDE) for Roadmapping of Emerging Technologies. Paper Presented at the Portland International Center for Management of Engineering and Technology, PICMET (2005)

Gerdsri, N.: An Analytical Approach to Building a Technology Development Envelope (TDE) for Roadmapping of Emerging Technologies. International Journal of Innovation & Technology Management 4(2), 121–135 (2007)

Gerdsri, N., Assakul, P., Vatananan, R.S.: Applying change management approach to guide the implementation of technology roadmapping (TRM). Paper Presented at the Portland International Center for Management of Engineering and Technology (PICMET), Cape Town, South Africa (2008)

Gerdsri, N., Assakul, P., Vatananan, R.S.: An activity guideline for technology roadmapping implementation. Technology Analysis & Strategic Management 22(2), 229–242 (2010)

Gerdsri, N., Dansamasatid, S.: Applying Technology Roadmapping to Drive the Strategic Medium-term Plan into Action: Lessons from the SCG Building Material Co. Sasin Journal of Management 14(1), 41–53 (2008)

Gerdsri, N., Vatananan, R.S.: Dynamics of Technology Roadmapping (TRM) Implementation. Paper Presented at the Portland International Center for Management of Engineering and Technology, PICMET (2007)

Gerdsri, N., Vatananan, R.S., Dansamasatid, S.: Dealing with the dynamics of technology roadmapping implementation: A case study. Technological Forecasting and Social Change 76(1), 50–60 (2009)

Groenveld, P.: Roadmapping integrates business and technology. Research Technology Management 40(5), 48 (1997)

Groenveld, P.: Roadmapping integrates business and technology. Research Technology Management 50(6), 49–58 (2007)

Hiatt, J.M.: ADKAR: A Model for Change Business, Government and our Community. Prosci Research Center, Loveland (2006)

Hoffman, M., Daim, T.U.: Building Energy Efficiency Technology Roadmaps: A Case of Bonneville Power Administration (BPA). Paper Presented at the Portland International Center for Management of Engineering and Technology, PICMET (2006)

Holmes, C., Ferrill, M.: The application of operation and technology roadmapping to aid Singaporean SMEs identify and select emerging technologies. Technological Forecasting and Social Change 72(3), 349–357 (2005)

Hosley, S.M., Lau, A.T.W., Levy, F.K., Tan, D.S.K.: The quest for the competitive learning organization. Management Decision 32(6), 5–15 (1994)

Howell, J.M., Higgins, C.A.: Champions of change. Business Quarterly 54(4), 31 (1990)

Kappel, T.A.: Perspectives on roadmaps: how organizations talk about the future. Journal of Product Innovation Management 18(1), 39–50 (2001)

Kostoff, R.N., Schaller, R.R.: Science and technology roadmaps. IEEE Transactions on Engineering Management 48(2), 132–143 (2001)

Kotter, J.P.: Leading Change: Why Transformation Efforts Fail. Harvard Business Review 73(2), 59–67 (1995)

Kotter, J.P.: Leading Change. Harvard Business School Press, Cambridge (1996)

Lee, S., Park, Y.: Customization of technology roadmaps according to roadmapping purposes: Overall process and detailed modules. Technological Forecasting and Social Change 72(5), 567–583 (2005)

McCall, M.W., Kaplan, R.E.: Whatever It Takes: Decision Makers at Work. Prentice-Hall, Upper Sawmill River (1985)

McMillan, A.: Roadmapping - Agent of change. Research Technology Management 46(2), 40–47 (2003)

Nonaka, I.: The Knowledge-Creating Company. Harvard Business Review 69(6), 96–104 (1991)

Phaal, R., Farrukh, C.J.P., Mitchell, R., Probert, D.R.: Starting-Up Roadmapping Fast. Research Technology Management 46(2), 52–58 (2003a)

Phaal, R., Farrukh, C.J.P., Mitchell, R., Probert, D.R.: Starting-Up Roadmapping Fast. IEEE Engineering Management Review 31(3), 54–60 (2003b)

Phaal, R., Farrukh, C.J.P., Probert, D.R.: Characterisation of technology roadmaps: purpose and format. Paper Presented at the Portland International Center for Management of Engineering and Technology (PICMET), Portland, OR (2001)

Phaal, R., Farrukh, C.J.P., Probert, D.R.: Technology roadmapping - A planning framework for evolution and revolution. Technological Forecasting and Social Change 71(1-2), 5–26 (2004)

Phaal, R., Farrukh, C.J.P., Probert, D.R.: Developing a technology roadmapping system. Paper Presented at the Portland International Center for Management of Engineering and Technology (PICMET), Portland, OR (2005)

Phaal, R., Farrukh, C.J.P., Probert, D.R.: Strategic Roadmapping: A Workshop-based Approach for Identifying and Exploring Strategic Issues and Opportunities. Engineering Management Journal 19(1), 3 (2007)

Probert, D., Shehabuddeen, N.: Technology road mapping: the issues of managing technology change. International Journal of Technology Management 17(6), 646–661 (1999)

Probert, D.R., Farrukh, C.J.P., Phaal, R.: Technology roadmapping - developing a practical approach for linking resources to strategic goals. Journal of Engineering Manufacture - Proceedings of the Institution of Mechanical Engineers Part B 217(9), 1183–1195 (2003)

Rinne, M., Gerdsri, N.: Technology Roadmaps: Unlocking the Potential of a Field. Paper Presented at the Portland International Center for Management of Engineering and Technology (PICMET), Portland, OR (2003)

Senge, P.M.: The fifth dimension: The Art & Practice of the Learning Organization. Random House, New York (1990)

Strauss, J.D., Radnor, M.: Roadmapping For Dynamic and Uncertain Environments. Research Technology Management 47(2), 51–57 (2004)

Vojak, B.A., Chambers, F.A.: Roadmapping disruptive technical threats and opportunities in complex, technology-based subsystems: The SAILS methodology. Technological Forecasting and Social Change 71(1-2), 121–139 (2004)

Author

Nathasit Gerdsri Ph.D. is an associate professor of technology and innovation management at College of Management, Mahidol University (Thailand). Currently, his research works in the area of technology roadmapping (TRM) have been focused on how to operationalize technology roadmapping process as well as how to effectively implement it in an organization. His research activities are carried out through academic and consulting projects. He received Ph.D. from Dept. of Engineering and Technology Management, Portland State U. A part of his dissertation on the development of technology development envelope (TDE) for roadmapping of emerging technologies was nominated to receive the outstanding paper award at PICMET in 2005. Prior to these, he worked for Intel's R&D lab in Hillsboro, Oregon.

Innovation Business Plan at Siemens: Portfolio-Based Roadmapping to Focus on Promising Innovation Projects Right from the Beginning

Babak Farrokhzad, Claus Kern, and Meike de Vries

Roadmapping is an effective tool for supporting innovation projects and business strategies. It is easy to implement and can be used in many different ways. In an international and globally operating company like Siemens, roadmapping offers senior management a valuable aid for decision-making that is easy to understand in any language. Siemens has developed a supportive approach to decision-making known as the innovation business plan. The core of this innovation business plan consists of a portfolio-based roadmapping process. The purpose of this paper is to provide an overview of the above-mentioned portfolio-based roadmapping approach within the innovation business plan and its compilation in the organization.

1 Introduction

The business environment in which Siemens operates is characterized by challenges of unknown dimensions due to demographic changes, urbanization, climate change and globalization. In addition, increasing demands by customers

---

Babak Farrokhzad
HOERBIGER Automatisierungstechnik Holding GmbH, Südliche Römerstraße 15, 86972 Altenstadt, Germany
e-mail: babak.farrokhzad@hoerbiger.com

Claus Kern
Siemens AG, E TI IM Technology & Innovation - Innovation Management, Freyeslebenstr. 1, 91058 Erlangen, Germany
e-mail: claus.kern@siemens.com

Meike de Vries
Competence Center Innovations- und Technologie-Management und Vorausschau, Fraunhofer-Institut für System- und Innovationsforschung ISI, Breslauer Straße 48, 76139 Karlsruhe, Germany
e-mail: meike.de.vries@isi.fraunhofer.de

result in ever shorter innovation and product life cycles, accompanied by rapidly evolving technologies. Economic and financial disruptions aggravate these challenges even more. For this reason Siemens possesses a very complex business structure comprising the four sectors of Industry, Energy, Healthcare, and the new sector Infrastructure & Cities. Each sector is organized in several divisions which in turn are divided into multiple business units operating all around the globe. Siemens looks back on a more than 160-year history of innovation, starting with the invention of the pointer telegraph in 1847, the discovery of the dynamoelectric principle by Werner von Siemens in 1866, via trendsetting industrial automation products such as SIMATIC. Recent highlights include the high density motor for electric cars, the offshore direct drive wind turbine, world class most efficient gas turbine 8000H, and the hydrogen electrolyser which transforms excess renewable energy into the easily storable chemical energy carrier hydrogen.

In order to maintain and expand its leading position as an innovative company, Siemens needs to identify particularly promising projects at an early stage of the innovation process, to direct its activities towards attractive market segments and to fit new developments into the business strategy. Siemens has implemented a variety of tools to achieve these goals, and summarized them in the concept of the "innovation business plan". In order to take into account foreseeable trends as well as possible discontinuities, Siemens also developed a process known as "Pictures of the Future". This visionary process combines two antinomic perspectives: extrapolation from "today's world" and retropolation from "tomorrow's world." The look ahead, extrapolation, corresponds to what companies usually do: mapping the development of the presently known technologies into the future and estimating as precisely as possible at what point in time something will be available. The retropolation as a scenario analysis completes the "Pictures of the Future" process. For a chosen time horizon, a scenario is designed that takes into account all of the influencing factors: the development of political and social structures, environmental impacts, technological trends, and new customer needs. This makes it possible to think one's way back into the present and identify tasks that must be accomplished in order to cope with this particular world of tomorrow. If both of these perspectives are reconciled in the "Pictures of the Future" process, consistent and internally coherent scenarios of the future are generated (Eberl, 2011).

In academic literature Siemens is well recognized and acknowledged as a global player and pioneer in the use of roadmapping (Specht and Behrens, 2006), having already implemented this technique in the early 1990s (Weyrich, 1996). By means of roadmapping it becomes possible to forecast developments based on present businesses and customers, and establish projections regarding the staging, timing and interplay of anticipated evolutions in product, process or infrastructure technologies (Phaal et al., 2003; Cosner et al., 2007). Roadmaps require a specific process to enable the forecast of future technologies' content and timing (Phaal et al., 2001; Möhrle, 2004). Hence, to become a powerful tool in the technological planning process, roadmapping must be linked with other processes, such as the long-range planning process and annual operating plan development (Cosner et al., 2007; Möhrle, 2004). Accordingly, Siemens has compiled a set of various methods to be combined in the innovation business plan concept, and defined the underlying process, including steps, roles, milestones and templates (Anton and Schnorr, 2006). The methodology of the innovation business plan will be introduced in the following section, and the key tool of portfolio-based roadmapping will subsequently be described in greater detail.

2 Methodology of the Innovation Business Plan

As a supportive method for decision-making in the annual business planning process, the innovation business plan is particularly suitable for evaluating current and yet-to-be-approved innovation projects, as well as for drawing up innovation plans and budgets for the following business year. Based on a well-defined and regularly updated process, the innovation business plan helps to improve the company's overall evaluation and selection abilities and enhances innovation performance. The innovation business plan consists of two major parts and possesses a pyramidal structure (Figure 1):

- The first part forms the basis of the innovation business plan; it is descriptive in character. In this part the actual business and technology environment is analyzed by means of consolidated market, technology and customer data. This consolidated data is graphically represented by a radar screen or strength-weakness profiles derived from the comparison with competitors. In the next step, this data is submitted to a minimum of three portfolio analyses and then condensed into a SWOT analysis. These assessments are supported and substantiated methodologically and are based on the preceding portfolio analyses. Recommendations are then formulated on the basis of these predominantly analytical descriptions. Thus, this part of the process primarily focuses on the preparation of innovation and R&D decisions.
- The second part centers on the decision making process and the visualization of these decisions in a roadmap. It forms the peak of the innovation business plan and is more general in character. Potential innovation projects are visualized in roadmaps based on the prepared data and the recommendations from part one. The derived roadmaps aid managers in decision making. Compiling the information in this manner provides a clear and comprehensible overview of all potential innovation projects, their position relative to the business strategy and the analyses concerning the business, competitor and technology environment.

The two-part structure of the innovation business plan, with its division into analysis and interpretation on the one hand and consequences on the other, has proven successful in business practice. Managers are in need of carefully compiled data relating directly to the entire business, as their decisions define action plans within the innovation business plan. However, Sales and Marketing, R&D managers, product portfolio managers and product managers involved in the preparation of decisions have a specific focus driven by their tasks when doing analyses. Therefore, the Technology & Innovation department compiles and aggregates these data and produces an impartial overview for the decision makers thus closing this gap.

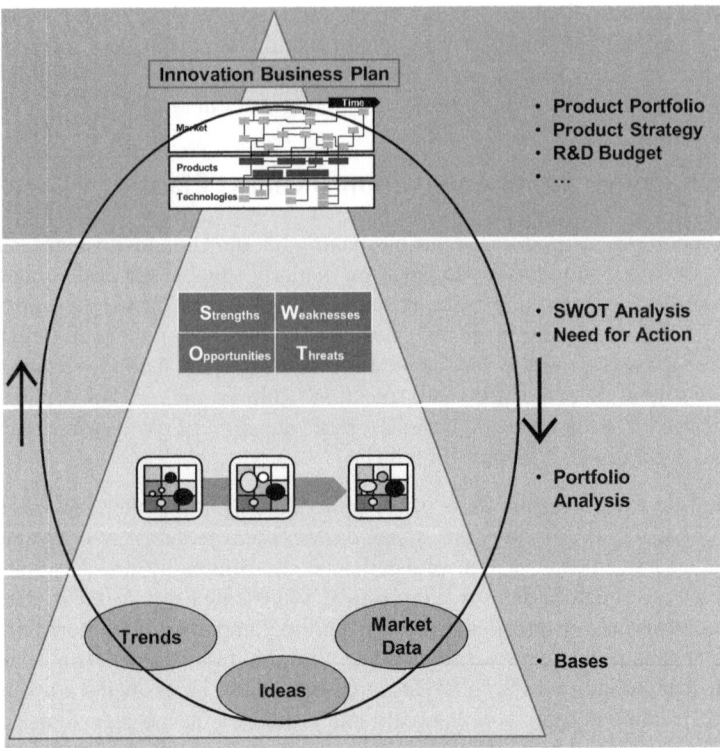

Fig. 1 Innovation business plan - basic structure and iterative cycle

The innovation business plan consequently contains concrete statements on:

- all innovation plans in a single list, including their budget and time requirements,
- a product and service roadmap,
- a technology and competence roadmap,
- a replacement strategy with a list of products to be phased out,

and ensures that these are developed systematically in the context of the business strategy.

3 Portfolio-Based Roadmapping as a Methodological Basis of the Innovation Business Plan

Siemens utilizes the innovation business plan to try and determine the timing of market entry for a technological solution as precisely as possible, or predict when a specific customer need is going to arise. The roadmapping approach is, above all, a visualization tool for innovation projects and therefore represents a substantial resource for decision-making on the senior management level (Farrokhzad et al., 2008). Siemens sees an effective tool in the combination of roadmapping and portfolio management, as portfolio analysis and roadmaps are easy to realize in the course of day-to-day business. Other authors also consider the combination of roadmapping and portfolio management a promising option (Cosner et al., 2007). The following section deals with the basis of portfolio management, the subsequent development of roadmaps, and finally the aspect of how these mechanisms are embedded in the organisation and implemented.

3.1 Portfolio Analysis

What a company needs to achieve short-term as well as sustained success is an appropriate mix of innovation projects (Kleinschmidt et al., 1996). The principal method for creating and managing the mix of innovation projects is that of portfolio management (Möhrle, 1999). In the case of Siemens this portfolio management encompasses a strategic and an operational process (Figure 2).

The strategic process, which is known as the product portfolio management (PPM) process, defines the product portfolio, the innovation projects and the evaluation of the contribution of innovation projects to the successful implementation of the company's business strategy. The strategic process is a rolling sequence of operations embedded into the Siemens process framework. The Siemens process framework includes all principles and regulations for process management within the company. Its core element is represented by the Siemens reference process house, which comprises the product lifecycle management process as one of its business processes. The front-end of the product lifecycle management process chiefly consists of the product portfolio management process. Product lifecycle management translates the business strategy as developed during the strategic planning and controlling process into business objectives for product lifecycle management and customer relationship management processes (Anton and Schnorr, 2006).

The operational process, which is known as the product management process (PMP), involves the design, definition, realization, marketing and eventual phasing out of products and services. The operational process is a linear process going on between a defined beginning and end.

The strategic product portfolio management process controls the operational product management process and thus the processing of innovation activities within the framework of the product management process. In the simplest case, the portfolio management process comprises three consecutive steps:

- evaluation of market data
- evaluation of technology data
- definition of product portfolio strategy and innovation projects.

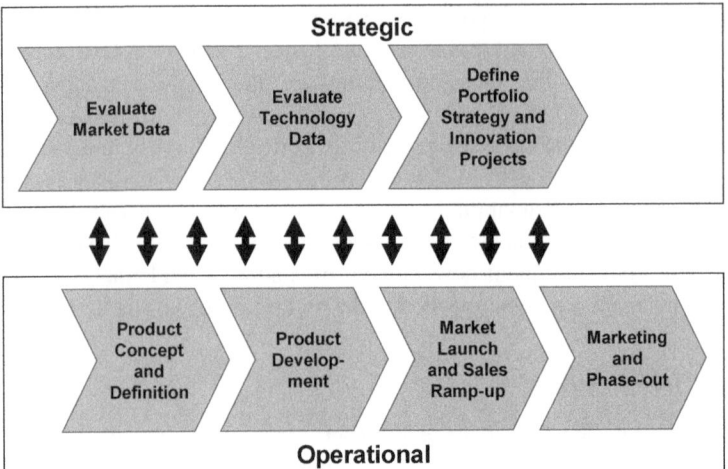

Fig. 2 Strategic and operative subprocesses in the overall product lifecycle management process

Each of these three steps consists of numerous individual operations. The process also includes a quantity of further steps that merely are of secondary importance in the systematic creation of roadmaps.

Producing a comprehensive analysis of all relevant information involves the consideration of both external and internal information sources, so portfolio analysis is an eminently suitable method (Specht et al., 2002). However, the popular market portfolios, such as those by McKinsey, obviously concentrate on the market cycle of a product, while the underlying technologies and developments are only considered indirectly (Specht et al., 2002; Steinmann and Schreyögg, 2005). Consequently, a market portfolio is unable to provide any indication of new strategic fields for the future. Market portfolios are only suitable for static environments.

Technology portfolios are a much better solution for dynamic, innovation-orientated businesses as they also take account of technological factors (Wolfrum, 1994). The aim of technology portfolios lies in the visualization of the technologies used in a product or company by means of a two-dimensional matrix for directing strategies towards future developments (Pfeiffer, 1991). But even though a technology portfolio is able to provide good indications of technology priorities, it yields no information concerning market priorities (Gerpott, 2005). A healthy and balanced strategic innovation management approach takes both technological and market success factors into account. The integrated market-technology portfolio, which includes technology planning and overall business planning, was devised to meet this particular need (Wolfrum, 1994).

Innovation Business Plan at Siemens: Portfolio-Based Roadmapping to Focus

The McKinsey approach generally comprises a simple technology portfolio and a market portfolio, which can then be combined in an integrated portfolio (Wolfrum, 1994; Gerpott, 2005). A classical portfolio analysis consequently consists of three consecutive steps:

- creation of a market portfolio
- creation of a technology portfolio
- creation of an integrated market-technology portfolio (Figure 3).

The market portfolio addresses market attractiveness and the relative market position, organizing products or product groups according to the company's market attractiveness and competitive strength in the respective market.

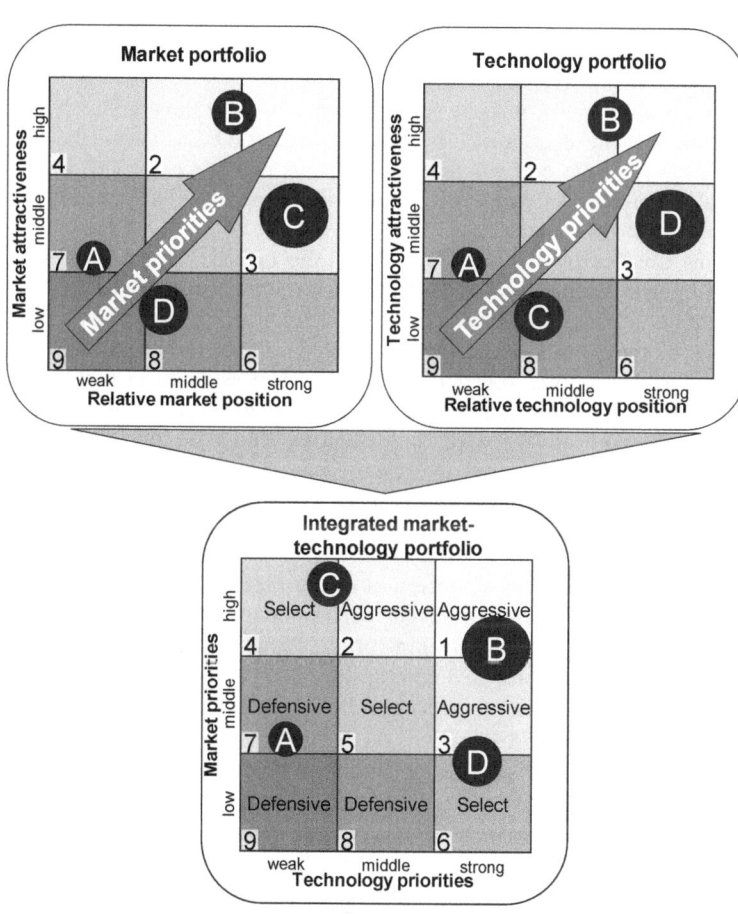

Fig. 3 Innovation business plan - core of portfolio analysis methodology

The technology portfolio addresses technology attractiveness and the relative technology position. In essence, the attractiveness of a technology depends on its position on the S-curve (Gerpott, 2005; Foster, 1986). The S-curve is a tool for the description of the market penetration of a product or service. It shows how much development potential still exists and how much costs of promoting will be needed. The relative technology position is based on the company's know-how background, as compared with that of its competitors, and the relative cost of promoting the technology. The order of technologies in the matrix is such that those marked by a higher degree of attractiveness and a higher relative position are assigned a higher priority.

The integrated market-technology portfolio, proposed first already in the beginning of the 90s, is a combination of market and technology portfolios with the purpose to address market priorities (Roussel et al., 1991). Technology-related priorities for R&D are established on the basis of the technology's position in the portfolio. At this point, a distinction is drawn between three model R&D investment strategies - i.e. aggressive, selective and defensive - which also indicate the strategic direction of impact.

As shown in Figure 3, it makes more sense to look at the combined portfolios than at each of them separately. For instance, product 'D' has the weakest position in the market portfolio and would gain only minor resources on this basis, but it has actually a high priority in terms of technology and would hence be best served by the selective R&D strategy.

Siemens uses an innovation project portfolio or risk portfolio similar to the McKinsey portfolio approach. This portfolio, which considers potential yield and the probability of success, allows the relative attractiveness of different innovation projects to be clearly illustrated by a comparison of opportunities and risks. The risk portfolio leads to a broader discussion of possible decisions than other portfolios: Decisions are not solely based on market potential, technological potential or the company's market and technology position, but also on the risks attached to innovation projects.

Siemens distinguishes a range of advantages in the portfolio-based roadmapping approach:

- The method enables a separation of the preparatory work carried out by experts in the various relevant departments preliminary to the decision from the actual decision-making process finalized by senior management.
- The method enables a demonstration of the R&D budget's impact on different innovation projects according to an ex-post analysis.
- The method supports decisions on just-in-time market entry or exit.
- The method facilitates the preparation of consistent roadmaps based not only on the business strategy, but also on an analysis of competition.
- The method helps to integrate and balance market and technology aspects.
- The method helps to integrate portfolio management and roadmapping into annual strategic business planning, which causes innovation activities to be more strategy-driven.

3.2 Roadmapping

The roadmaps originate from the above-mentioned portfolio analyses, which are respectively based on several other types of analysis. Siemens uses roadmapping to support decision-making at senior management level. A general distinction can be drawn between the core technology roadmaps on corporate level, which possess a medium- to long-term focus, and the product and technology roadmaps of various business divisions, which are marked by a short- to long-term focus.

Business division roadmaps are updated annually in a continuous process. Business division roadmaps are also referred to as multi-generation product plans or MGPPs. The various divisions' multi-generation product plans are aligned with the core technology roadmaps on corporate level. This step serves as a check on competencies and the roadmaps' consistency (Figure 4) (Weyrich, 1996).

Siemens, for example, features several core technology roadmaps and several hundreds of product and technology roadmaps. In this context it is noteworthy that the relatively small quantity of core technology roadmaps significantly heightens the acceptance of roadmapping across the company and among employees (Bucher, 2003).

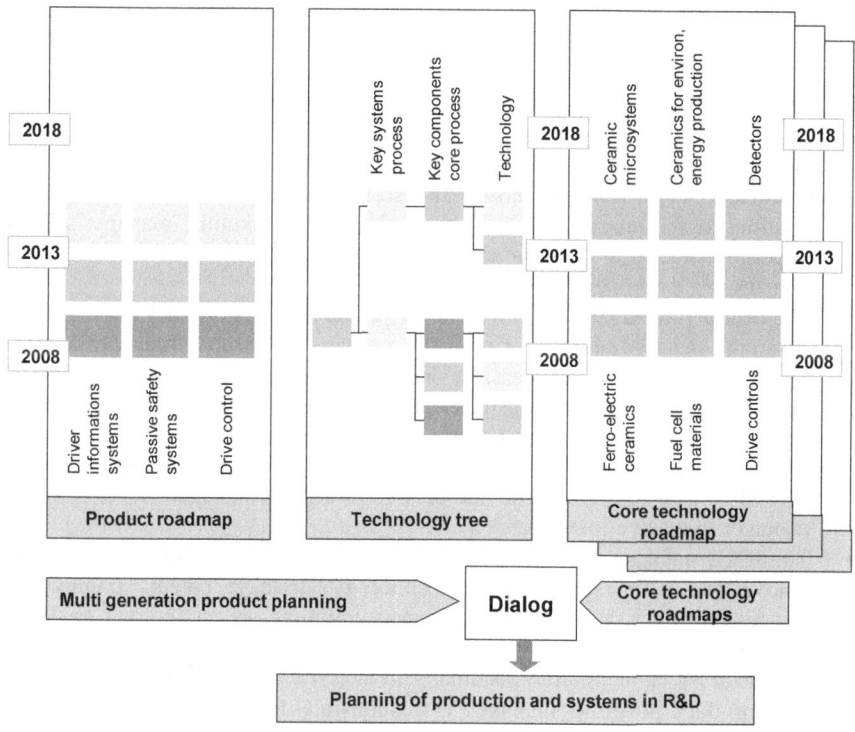

Fig. 4 Key elements of the innovation business plan

3.3 Embedding into and Implementation in the Organization

The portfolio-based roadmapping methods and organization have to be implemented in accordance with company structures if the scheme is to tie up successfully with the innovation business plan (Farrokhzad et al., 2008). When a new method needs to be implemented in a company, the respective process should be specifically defined.

The first step of embedding portfolio-based roadmapping in the organization is to generate a precise delineation of the procedure using defined processes (Phaal and Farrukh, 2000; Groenveld, 2007). The distribution of roles both defines the primary functional areas and staff involved in producing the innovation business plan, and describes how their specific perspectives within the company are to be merged into an innovation strategy and systematically converted into a roadmap.

Siemens has defined e.g. five roles within the company that are essential to the success of an innovation business plan:

- sales and marketing manager,
- product manager,
- R&D manager,
- product portfolio manager,
- business manager.

- The sales and marketing manager represents the customer and sales point of view in product portfolio management and contributes the strategic market perspective, which may for example include market trends, customer characteristics, strengths and weaknesses, and opportunities and risks in relation to the market. One of the sales and marketing manager's most important tasks is to open up and develop sources of information with respect to market and competition data.
- The product manager is responsible for positioning the product in the market and represent the interests of relevant products within the framework of the product management process. It is his or her responsibility to identify problems affecting the business that arise in the course of product development or product marketing and to ensure that these are resolved effectively. Depending on the nature of the problem, the product manager may also have to confer with the product portfolio manager or the business manager to reach a joint decision.
- The R&D manager is entrusted with the planning and the implementation of innovations with regard to the technology strategy, patent strategy and standardization strategy. He or she also commissions technology analyses, initiates analyses of core competencies from a technological point of view, and ensures that technology roadmaps are derived from portfolios.
- The product portfolio manager is responsible for the positioning of the product portfolio in the market from technical, economic and strategic points of view. His or her actions are based on the business strategy as formulated by the senior management. The product portfolio manager is the process owner

of all processes involved in the correct performance of overall product portfolio management, including the requisite analyses. Hence it is important that the product portfolio manager reports directly to the business manager, so he or she can act regardless of divergent interests that may exist in the company's functional areas.

- The business manager formulates the business strategies and targets, which have to be in line with the overriding strategies of the company as a whole. He or she approves of or adjusts the recommended actions for the innovation portfolio that have resulting from the product portfolio management process. He or she also appoints a product portfolio manager to the relevant management unit, assigns responsibility for product development and selects project leaders for concrete innovation projects.

The various roles that are necessary in the framework of an innovation business plan determine what specific influence certain individuals in the company should have. This also involves a detailed description of the tasks allotted to different roles. The interfaces between the three vertical roles of business managers, sales and marketing, and R&D, plus the two functional roles of product portfolio manager and product manager can be defined on this basis.

The process should be simultaneous with the annual planning and budgeting processes. This ensures that the link between business strategy and planning and innovation management is maintained and facilitates a much closer alignment of business strategy and innovation activities.

4 Conclusions

This paper provides a presentation of the innovation business plan as used by Siemens. The respective concept is, first and foremost, a tool to support decision making processes in the company, helping to ensure that taken choices result in an appropriate innovation mix and appropriate innovation projects or strategies. The innovation business plan consists of clearly defined methodologies, processes and roles. The integration of this process into annual planning helps strengthening the link between innovation management and business strategy. Strong analytical capabilities and strong leadership by management are factors which have been identified as being absolutely vital for success.

References

Anton, O., Schnorr, J.: Der Schlüssel zur kontinuierlichen Innovation heißt Siemens PLM. CADplus Business + Engineering 9(6), 4–6 (2006)

Bucher, P.E.: Integrated Technology Roadmapping: Design and Implementation for Technology-based Multinational Enterprises. Dissertation Thesis. Zurich: Swiss Federal Institute of Technology (2003)

Cosner, R.R., Hynds, J., Fusfeld, A.R., Loweth, C.V., Scouten, C., Albright, R.: Integrating roadmapping into technical planning. Research Technology Management 50(6), 31–48 (2007)

Eberl, U.: Life in 2050. Beltz & Gelberg, Weinheim (2011)

Farrokhzad, B., Kern, C., Fritzhanns, T.: Innovation Business Plan im Hause Siemens - Portfolio-basiertes Roadmapping zur Ableitung Erfolg versprechender Innovationsprojekte. In: Möhrle, M.G., Isenmann, R. (eds.) Technologie-Roadmapping. Zukunftsstrategien für Technologieunternehmen, pp. 325–352. Springer, Berlin (2008)

Foster, R.N.: Innovation: The Attacker's Advantage. Summit Books, London (1986)

Gerpott, T.J.: Strategisches Technologie- und Innovationsmanagement. Schäffer-Poeschel, Stuttgart (2005)

Groenveld, P.: Roadmapping integrates business and technology. Research Technology Management 50(6), 49–58 (2007)

Kleinschmidt, E.J., Geschka, H., Cooper, R.G.: Erfolgsfaktor Markt. Kundenorientierte Produktinnovation. Springer, Berlin (1996)

Roussel, P.A., Saad, K.N., Tiby, C.: Management der F&E-Strategie. Gabler, Wiesbaden (1991)

Möhrle, M.G.: Der richtige Projekt-Mix. Springer, Berlin (1999)

Möhrle, M.G.: TRIZ-based technology-roadmapping. Int. J. Technology Intelligence and Planning 1(1), 87–99 (2004)

Pfeiffer, W.: Technologie-Portfolio zum Management Strategischer Zukunftsgeschäftsfelder. Vandenhoeck & Ruprecht, Göttingen (1991)

Phaal, R., Farrukh, C.: Technology Planning Survey - Results. Centre for Technology Management, Project Report, University of Cambridge, March 14 (2000)

Phaal, R., Farrukh, C., Probert, D.: Technology Roadmapping: Linking Technology Resources to Business Objectives (2001), http://www.ifm.eng.cam.ac.uk/ctm/publications/tplan/trm_white_paper.pdf (accessed July 11, 2011)

Phaal, R., Farrukh, C., Mitchell, R., Probert, D.: Starting-up roadmapping fast. Research Technology Management 46(2), 52–58 (2003)

Specht, D., Behrens, S.: Produkt- und Technologieroadmapping (2006), http://heilbronn.ihk.de/upload_dokumente/infothek/anlagen/7392_2532.pdf (accessed November 01, 2006)

Specht, G., Beckmann, C., Amelingmeyer, J.: F&E-Management. Kompetenz im Innovationsmanagement. Stuttgart: Schäffer-Poeschel (2002)

Steinmann, H., Schreyögg, G.: Management, Grundlagen der Unternehmensführung: Konzepte, Funktionen, Fallstudien. Gabler, Wiesbaden (2005)

Weyrich, C.: Zentrale Forschung und Bereichsentwicklung als Speerspitze der internationalen Innovationsaktivitäten bei Siemens. In: Gassmann, O., Zedewitz, M.V. (eds.) Internationales Innovationsmanagement: Gestaltung von Innovationsprozessen im Globalen Wettbewerb, pp. 119–125. Vahlen, München (1996)

Wolfrum, B.: Strategisches Technologiemanagement. Gabler, Wiesbaden (1994)

Authors

Babak Farrokhzad is Executive Vice President at Hoerbiger Automation Technology where he is responsible for Business Development and Innovation. Under his leadership, the organization has identified a series of intelligent electro-fluidic actuators with trendsetter potential that are under development of market introduction. Before, he was Director of Innovation Management in the Siemens Automation & Drives Group responsible for implementing methodologies that specifically support radical and disruptive innovations, and for developing appropriate trendsetting strategies. After receiving his Ph.D. in electrical engineering, Dr. Farrokhzad started his career with Roland Berger Strategy Consultants and joined later at the Corporate Strategy Department of Infineon Technologies and the Corporate Technology Department of Siemens.

Claus Kern has been Innovation Manager since 2001. Today he is Head of Innovation Management in the Siemens Energy Sector and amongst others responsible for the sector-wide innovation management, particularly for the systematic formulation of the innovation and product strategy. After studying Telecommunication Engineering, Claus Kern started out in the development of real-time systems for managing energy transmission and distribution grids. After that he worked as a Director for customer projects and later as a Director for product management in the field of energy automation and power quality. From 2001 to 2008 he worked as Innovation Manager in the overall field of Power Transmission and Distribution. Since 2008 he is working in the Energy Sector and 2010 he was assigned as "Principal Key Expert Energy Systems".

Meike de Vries is a Research Associate at the Fraunhofer Institute for Systems and Innovation Research ISI in the Competence Center Innovation and Technology Management and Foresight. She completed her studies in business administration focusing on Technology and Innovation Management in 2007 and worked afterwards as a Research Associate at the Institute of Project Management and Innovation (IPMI) at the University of Bremen. There, her work focused on roadmapping and she is writing her doctoral thesis on "Innovation Communication with Roadmaps".

Exploratory Roadmapping: Capturing, Structuring and Presenting Innovation Insights

David A. Beeton, Robert Phaal, and David R. Probert

Exploratory roadmaps are created to enhance future outlook. This is typically a process of exploring and expanding the options available to investigate a strategic area. This paper outlines how a standardised exploratory roadmapping process can be applied to capture and structure insights from across supply chains and to develop future views of the competitive issues facing a diverse industrial area. The application of this process produced a roadmap that provided useful information, structure and context for strategic planning and innovation processes in a complex multi-stakeholder industry.

1 Introduction

The flexibility of the roadmapping approach is often cited as one of its greatest attributes. As a result there is considerable diversity among roadmapping efforts in terms of their objectives, techniques employed, output format and content. This paper outlines a standardised roadmapping process that was developed to structure data collection and analysis in a three year industrial project to produce the International Roadmap for Consumer Packaging (Beeton, 2006). This process is broadly based on the 'T-Plan' approach (Phaal et al., 2001) and was specifically designed to develop an exploratory roadmap for sector-level foresight.

2 Exploratory Roadmapping

Exploratory roadmaps are produced to enhance future outlook or foresight. This is typically a process to explore and expand the options available to develop

---

David A. Beeton
Urban Foresight Limited, 8 The Crescent, Newcastle upon Tyne, NE7 7ST, United Kingdom
e-mail: david.beeton@urbanforesight.org

Robert Phaal · David R. Probert
Centre for Technology Management, Institute for Manufacturing, Department of Engineering, University of Cambridge, 17 Charles Babbage Road, Cambridge, CB3 0FS, United Kingdom
e-mail: {rp108,drp1001}@cam.ac.uk

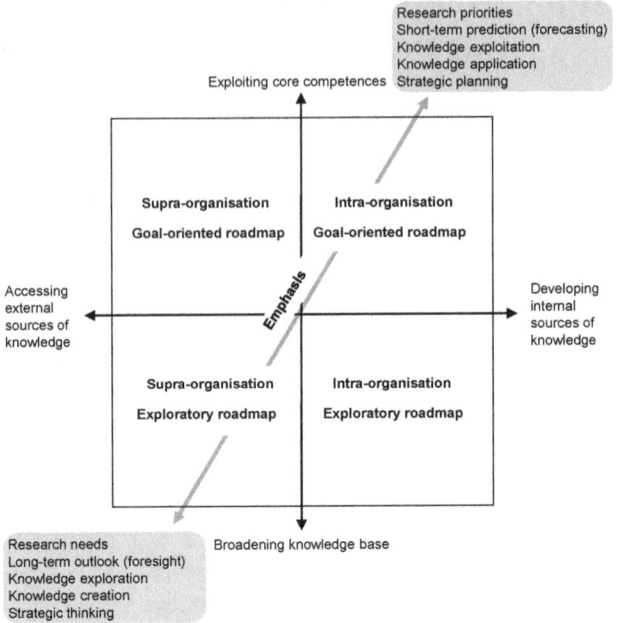

Fig. 1 An exploratory-goal-oriented roadmap taxonomy (Beeton, 2007)

understanding of an industrial landscape or a competitive position (Beeton et al., 2008). As shown in Figure 1, this is at the opposing end of the spectrum to convergent goal-oriented roadmaps that outline a sequence of activities and actions to define how a strategy may be implemented or an objective achieved. Also, as shown in Figure 1, exploratory roadmaps emphasise the broadening of an organisational knowledge base and therefore invariably engage multiple actors.

3 Developing an Exploratory Roadmap for Foresight

The International Roadmap for Consumer Packaging is the output of an industrial project to explore the research interests of the membership networks of two UK-based organisations (Pira International and the Faraday Packaging Partnership) and the wider packaging sector. This presented the opportunity for detailed investigation of the process of developing an exploratory roadmap and documentation of the key findings that emerged.

The explanatory scheme in Figure 2 was developed to structure these findings. This categorises the main activities and learning derived, explaining how insights were collected in a series of 12 roadmapping workshops and then processed to develop the final roadmap.

3.1 Planning

Starting up the roadmapping process and developing a robust approach can present significant challenges (Phaal et al., 2000). This can be a particular problem in developing sector-level roadmaps, where the diverse interests and needs of different stakeholders can create additional complexity. In developing the packaging roadmap much of this complexity was addressed in the planning stage, whereby a number of key activities were undertaken to accommodate these requirements and manage potential risks.

3.1.1 Establish a Steering Committee

An important first step in initiating the roadmapping activity was to establish a steering committee that was responsible for the design and implementation of the process. The key participants in this group were a process owner who provided expert-industry knowledge and acted as a champion for the roadmapping activity, and a process facilitator who was responsible for the running and co-ordination of workshops and the processing of the workshop outputs.

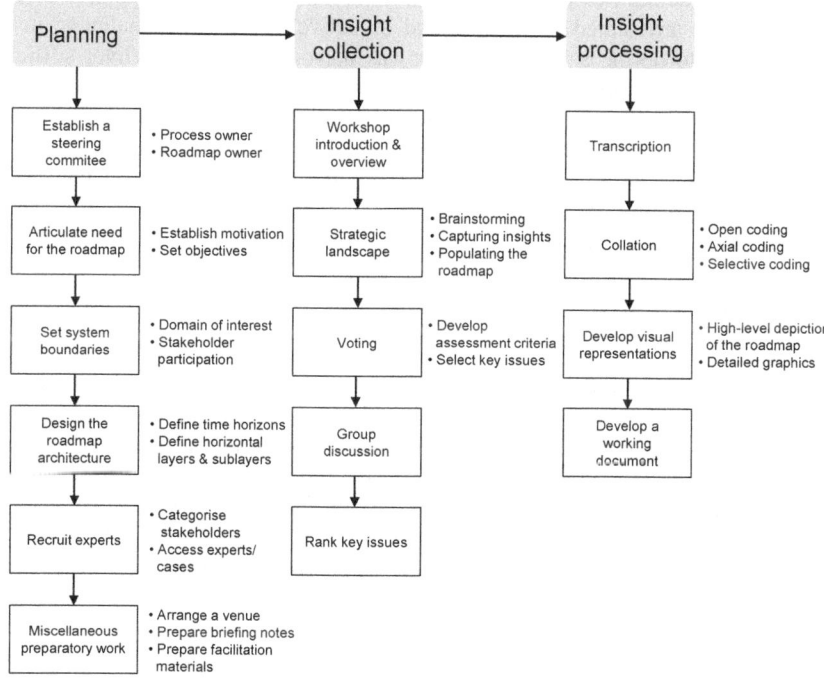

Fig. 2 The general features of the process of developing an exploratory roadmap

3.1.2 Articulate the Need for the Roadmap

A key responsibility of the steering committee was to establish the motivation for the roadmap and the future value that could be realised from this activity. The overarching aim in this case was to explore strategic opportunities and threats facing the packaging sector. Establishing this broad motivation characterised the approach as being more exploratory than goal-oriented. Therefore, the expected outcome was to sketch a landscape of general needs of the sector by establishing a list of things deemed to be important. This led to the definition of the following three main objectives:

- To capture and structure the key trends and drivers facing the packaging sector over the next 10 years.
- To communicate detailed insights into the nature and implications of these trends and drivers, including identification of competitive threats and opportunities for innovation.
- To provide a framework to support strategic planning, decision-making and collaboration in the packaging sector.

3.1.3 Scope

Defining the domain of interest was a key planning activity, which was directly related to the articulated motivation and objectives. A specific challenge in this process was to define a landscape that was narrow enough to deliver sufficient detail for the roadmap, but not restricted in a way that would stifle creativity and constrain insights to what was conventionally accepted in the sector.

A sectoral system approach was adopted whereby the boundaries were delineated by activities that were unified by related product groups and shared knowledge. Specifically, in this case, the system boundaries were set to only include organisations and experts who had knowledge related to consumer packaging.

3.1.4 Roadmap Architecture

The defined domain of interest also influenced the roadmap architecture, which both structured insight collection in roadmapping workshops and organised the information communicated in the final roadmap. The design of the roadmap architecture was based on the generic roadmap template of a time-based chart, with the temporal dimension on the horizontal axis and the vertical axis broken into a number of broad layers. The horizontal axis of the roadmap was set at a 10-year time horizon and was sub-divided into categories of short, medium and long-term. In addition to these horizons, the architecture categorised historical issues (i.e. events that have occurred in the past, but remained relevant) and issues that may occur beyond 10 years, which were classified as visions, predictions or aspirations.

The specific themes associated with the roadmapping activity were incorporated into the broad layers on the vertical axis of the roadmap. The generic roadmap illustrated in Figure 3 shows that this usually takes the form of four layer types. The first represents the market trends and drivers that influence the development of the packaging sector, which, as shown in Table 1, is divided into five sub-categories using STEEP factors (social, technological, environmental, economic and political). The second layer of a generic roadmap is a product layer. However, the notion of 'product' differs along packaging supply chains. For example a product to a paperboard manufacturer is a refined raw material, to a packaging converter it is the packaging itself (e.g. a carton) and to the brand owner or retailer it is an integrated system of a consumer good within a package (e.g. a filled carton). In addition, the fact that the system boundaries were drawn to include any company that had knowledge or interest related to packaging meant that the study also incorporated organisations that were not directly involved in manufacturing a consumer packaged product, such as raw materials suppliers, machinery manufacturers, designers and consultants. Therefore, to emphasise that the focus of the roadmap was packaging, this layer was given the title of 'packaging performance areas'. The sub-categories in this layer were derived from a model of the lifecycle of packaging.

The technology layer of the roadmap architecture specifically considers available and emerging technologies that are directly related to the packaging sector. This differentiates it from the technological sub-category of the market layer, which considers technologies external to the packaging sector. The technology layer was divided into two sub-categories of 'product' and 'process'. This is consistent with the Schumpeterian view of product and process innovation, which classifies a 'product' as a good or service offered to the customer or client and a 'process' as the mode of production and delivery of the good or service.

The fourth layer of the roadmap architecture considers the development of resources that are not attributable to the other three broad layers. These resources are notionally considered to support the development of technologies and products, incorporating considerations such as capital, finance, skills, partnerships and supply chain interactions.

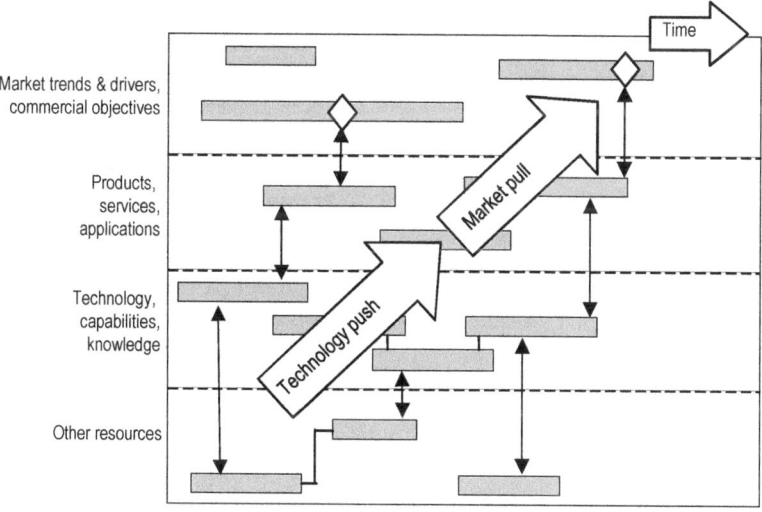

Fig. 3 A generic roadmap architecture

Table 1 Definitions developed for the layers in the roadmap architecture

Broad layer	Sub-categories	Outline definition
Market	Social	The social systems in which we live
	Technological	How technology external to the packaging sector affects the way that we live
	Environmental	The physical environment in which we live
	Economic	The global, national, corporate and personal financial systems that affect our lives
	Political	The systems that govern us including policy, regulation, legislation, and other political processes
	Other	Any information that does not conform to the above identified categories
Packaging performance	Materials and manufacturing	The requirements associated with the raw materials and manufacturing processes that serve to create a packaged product
	Transport and storage	The function of packaging in the distribution and storage of goods
	Retail/ transfer to consumer	The role of packaging in facilitating the sale/transfer of goods to the consumer, in addition to meeting other requirements of the retail environment
	Use by consumer	The specific requirements of packaging in the use of consumer products
	Recycling and disposal	The associated requirements placed on packaging with reference to both recycling and disposal
	Other	Any information that does not appear to conform to the above identified categories
Technology	Product	Technology inherent in products (i.e. packaging)
	Process	Technology associated with the systems and processes that create and interact with packaging over a lifecycle
	Other	Any information that does not appear to conform to the above identified categories
Resources		Underpinning resources to support the development of technologies and products.

3.1.5 Recruiting Experts

A key element in designing the roadmapping process was to identify the people with the knowledge and expertise necessary to develop a well-founded and credible roadmap. Furthermore an explicit objective was to collect insights from diverse stakeholders across the sector. To achieve this, the relative position of organisations in the supply chain was the key dimension used to categorise stakeholders. Geography was another variant, with workshops held in the UK, Germany and USA. The recruitment of these experts was made considerably easier by the existence of an established network of organisations who were affiliated to the process owner.

A number of additional activities were also performed prior to the workshops. These were mainly administrative and included formally inviting participants, booking an appropriate venue, preparing briefing documents and facilitation materials.

3.2 Insight Collection

Roadmapping is an expert-led processes, with actors typically convened in workshops. This provides an opportunity for interaction across different industry segments and organisations, providing a unique meeting place for people that may not normally come together.

Twelve workshops were held in total in this process, bringing together almost two hundred delegates representing seventy-one different organisations from across the packaging sector.

3.2.1 Workshop Format

The format of the workshop was proposed by the process owner and agreed with the roadmap owner. This was essentially a process of deciding on the activities that would take place in the workshop, identifying an appropriate sequence of events and allocating an amount of time to each activity.

Each workshop commenced with a formal presentation by the facilitators, which introduced the objectives of the workshop and the wider aims of the project. An overview of the roadmapping technique was provided, with an explanation of the processes to be used in the workshop.

3.2.2 Strategic Landscape

This activity used 'brainstorming' to identify strategic issues to characterise the range (or landscape) of competitive opportunities and threats facing the sector. Brainstorming was originally developed by Osborn (1963) who suggested four basic rules for a successful group brainstorming session, which were broadly followed in this process:

- Criticism is ruled out – adverse judgements of ideas must be withheld until later in the process;

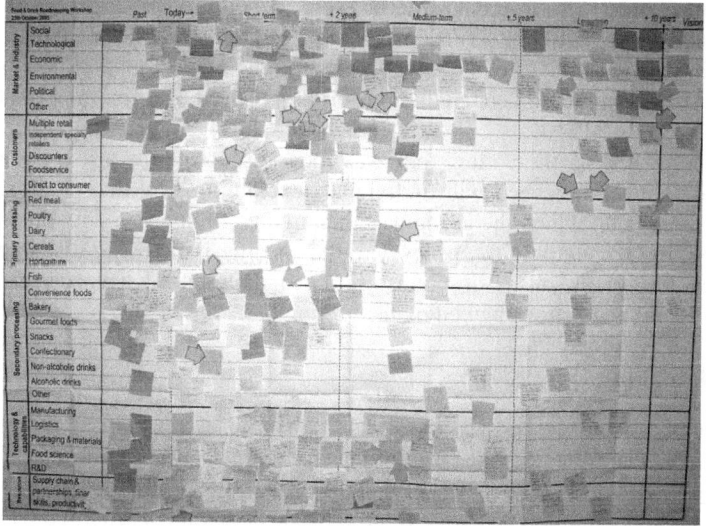

Fig. 4 An example of the output of a strategic landscape brainstorming activity

- "Free-wheeling" is welcomed – the wilder the idea, the better, as "it is easier to tame down than to think up";
- Quantity is wanted – the greater the number of ideas is more likely to produce useful ideas;
- Combination and improvement of ideas are sought.

Workshop participants were asked to write issues on sticky notes that they believed to be important. As shown in Figure 3, these comments were then placed on a large wall-mounted chart. No specific rules were suggested as to the content or format of the comments that were expected from the participants. However, outline definitions of each of the sub-layers of the roadmap were provided.

Two approaches were trialled in different workshops for the way in which the wall-mounted chart was populated with sticky notes. The first was an open approach where participants were invited to place comments in any of the layers of the roadmap and in any order. The second was a more disciplined approach where participants were asked to only consider one broad layer at a time. The main advantage of the 'open' approach was that it took less time. The layer-by-layer approach, however, was found to produce a significantly greater number of comments.

3.2.3 Voting

A voting process was used to identify key issues from the content of the strategic landscape. Workshop participants were asked to select the sticky note comment that they believed to the most important issue.

A set of assessment criteria was developed using group brainstorming to clarify the voting process. Different criteria were developed in each workshop to accommodate the different priorities of each group. It was decided that using a pre-defined set of criteria could have influenced the workshop participants in this process. Furthermore, understanding this criteria proved to be an important output of the process.

Immediately after the voting, time was spent discussing the priority issues that emerged. This commenced with individuals explaining why they had selected a specific comment, which the group was then invited to improve with the intention of generating a soundbite to encapsulate the specific issue being discussed. This followed Osborne's rules for brainstorming and produced a consensus list of priority issues and actions agreed by the group.

3.2.4 Ranking

The list of priority issues was ranked by giving each participant five stickers, which they were asked to place next to the five items that they believed to be most important. As with the voting process, individuals were asked to vote on the basis of their personal opinion and were referred to the assessment criteria developed in the workshop. The output of this process was a ranked list of priority issues and actions.

Exploratory Roadmapping: Capturing, Structuring and Presenting Innovation 233

3.3 Insight Processing

The information gathered in the twelve workshops was collated using a coding process to produce a series of categories under each of the layers on the roadmap. Sixty-four of these categories were identified in total to summarise the range of issues identified in the workshops. A particular advantage of this approach was that it provided a high-level summary of these issues, with an indication of the time-horizon in which the participants believed they were likely to impact the packaging sector.

3.3.1 Collation

The procedure used to structure analysis of the insights collected in the roadmapping workshops was open, axial and selective coding phases proposed by Strauss and Corbin (1990).

Open coding: Open coding is a process of identifying, naming and categorising the essential ideas found in the data. This can be divided into two phases: conceptualising and categorising. In conceptualising, incidents, ideas, events and acts are selected from the data and labelled by concepts. These codes either explicitly summarise the comment analysed, or are interpreted by the process owner. For example, in this roadmap, the comment "increasing proportion of the population is elderly" was considered explicit and coded as 'ageing population'. However, the meaning of the comment "we are a greying society" is more ambiguous, but was coded by the process owner as also referring to the ageing population.

Another key aspect of the coding process, of which examples are shown in Table 2, is that an individual comment may contain more than one code. In these instances, the comments were divided into separate codes. This was an emergent process, which demanded that codes were revisited several times as new patterns appeared.

The open coding process was also used to relate the individual codes to different time horizons. An example of how this is shown in Table 3, whereby codes were assigned to the time horizon in which they were most commonly placed in the strategic landscape brainstorming.

In the categorising phase of open coding, codes were grouped into more abstract, higher order categories. This enabled identification of common properties and analysis of what might occur if one of these properties were to change. For example, Table 4 shows the grouping of a number of concepts under the abstracted code of 'households.'

Table 2 Example of the process of analysis of the data collected in the strategic landscape brainstorming activities

Post-it note comment	Open code(s)
More single person households	Increase in single person households
Increase in single households increases the packaging use and increases the waste	Increase in single person households / increased consumption of packaging / increased production of waste
Natural limits to petroleum products - need to recycle	Consumption of finite resources / recycling of finite resources

Table 3 Placing an open code in an appropriate time horizon

Comment	Open code	Time horizon	Assigned time horizon
Internet shopping will become more popular	Internet shopping	Long	Medium
Widening use of internet shopping	Internet shopping	Medium	
Virtual supermarket	Internet shopping	Medium	
Increased use of internet shopping for all types of goods	Internet shopping	Short	
Increased use of home shopping (internet orders for groceries)?	Internet shopping	Medium	
Online shopping	Internet shopping	Medium	
The post-supermarket era (i.e. all internet based grocery shopping)	Internet shopping	Long	

Table 4 Example of how open codes were grouped into higher-order categories

Open code	Abstracted code
Working mothers Divorced parents Single parents Increase in single person households Increase in divorce Changing family structures Children staying in parental home longer	Households

Table 5 Example of how open and abstracted codes were related to the broad layers and sub-layers of the roadmap architecture

Open code	Abstracted code	Sub-layer	Broad layer
Increase in single person households Fewer people marrying Increase in divorce Changing family structures Children staying in parental home longer Inflated housing market	Households	Social	Market

An important part of this process was the naming of the categories. This process primarily drew on the data itself, whereby the most commonly occurring term for a category was used. For example, a number of different terms were used to describe the phenomenon of the increased average age of the general population (such as "elderly population" and "greying society"). The term that featured most often was "ageing population" and was therefore used to name this category. The names of these categories were validated by the roadmap owner to ensure consistency with the language used in the sector.

These abstracted codes were assigned to appropriate time horizons on the basis of the earliest corresponding open code. An arrow was used to illustrate that it was an ongoing trend where other corresponding open codes were categorised in different time horizons.

Exploratory Roadmapping: Capturing, Structuring and Presenting Innovation   235

Axial coding: Axial coding is a process of developing a deeper understanding of the relationships in the phenomena underlying data through the process of connecting various data categories. This is a process of obtaining an even higher level of abstraction and quantifying relationships. This was achieved by taking the codes developed in the categorising phase and relating them to the broad layers and sub-layers of the roadmap architecture. As shown in Table 5, this placed each of the categories in a defined hierarchical context and further enabled the identification of relationships between the data.

The roadmap consciously avoided being prescriptive. Hence relationships between data were only identified where reference was made in the data collected. The method used to do this was to re-examine the sticky note comments that contained more than one code to identify any explicit links. For example, as shown in Table 2, the comment "increase in single households increases the packaging use and increases the waste" was divided into three separate open codes. Hence, it was possible to identify a link between each of these codes and the higher-order abstracted codes in which they were categorised (i.e. increase in single person households → increased consumption of packaging → increased production of waste).

Selective coding: Selective coding is a process of developing the theory that best fits the phenomena by identifying a story that reveals the central phenomenon (the core issue or core category) under study. This was achieved by integrating the various codes and categories into the broad layers of the roadmap architecture to establish the following explanatory context:

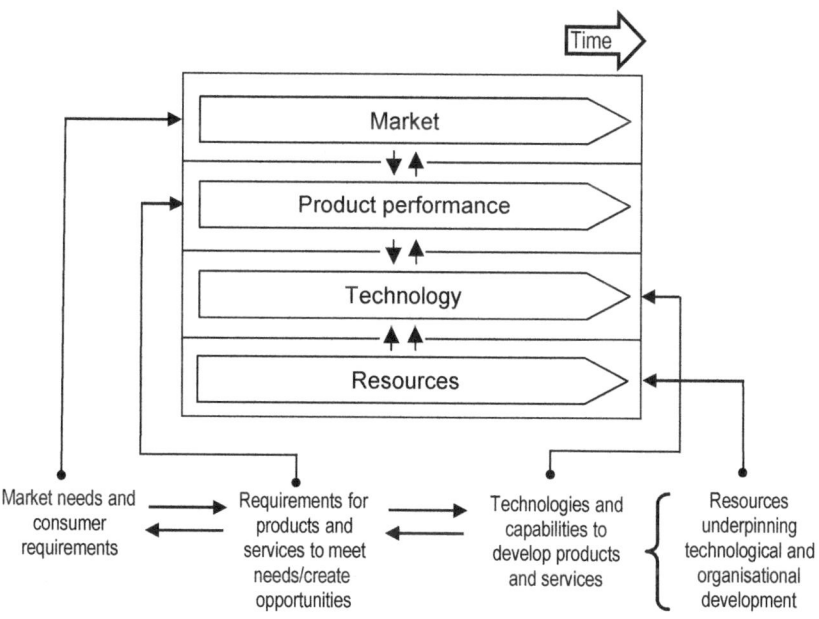

Fig. 5 The underlying logic of the broad layers of the roadmap architecture

- The market layer identifies evolving market requirements and business objectives.
- The technology layer depicts emerging technological requirements and capabilities.
- The product layer represents the tangible products and services that are developed to respond to these trends and drivers (i.e. a product is a technology with a market application).
- The resources layer underpins the technological capabilities and organisational requirements, identifying the resources that need to be in place to make development possible.

Developing a visual representation: A key objective in this process was to obtain a level of abstraction that enabled clear communication, whilst maintaining a sufficient level of detail. To facilitate this, a hierarchical approach was adopted, whereby the data was structured into two levels. The first graphic developed (shown in Figure 6) is a high-level depiction of the roadmap presenting the abstracted codes in the broad layers and sub-layers of the roadmap architecture. The key aspects of this illustration are:

- It provides a succinct summary of the range of issues covered in the roadmap;
- It provides a clear depiction of the categories of abstracted codes created by the broad layers and sub-layers of the roadmap architecture;
- It illustrates the time horizons in which individual abstracted themes are believed to be most important.

A detailed graphic was developed for each of the abstracted codes. An example of this is shown in Figure 7, which illustrates the open codes that were categorised into the abstracted code of 'ageing polulation'. The key aspects of these illustrations are:

- The broad layers and sub-layers of the roadmap architecture in which the abstracted code is placed are shown;
- Open codes are depicted in appropriate time horizons;
- Abstracted codes are shown as ongoing trends starting at the point of the earliest occurring open code (i.e. the earliest time horizon);
- The relationships between certain open codes are identified with arrows.
- Links to other abstracted codes are shown.

Where a single open code was related to other open codes, a single arrow was used. For example, as shown in Figure 7, the growth of the grey market may result in the average consumer having more or less disposable income. However, the number of arrows included in the illustration is kept to a minimum as it was found that a greater amount increased the complexity of the graphic. A further feature of these illustrations was to incorporate links to other abstracted codes. As can be seen in Figure 6, this is shown in the form of a box that refers the reader to a related section in the report. In this instance, a link has been made between the open code of 'more disposable income' and the headline theme of 'personal wealth'.

Exploratory Roadmapping: Capturing, Structuring and Presenting Innovation

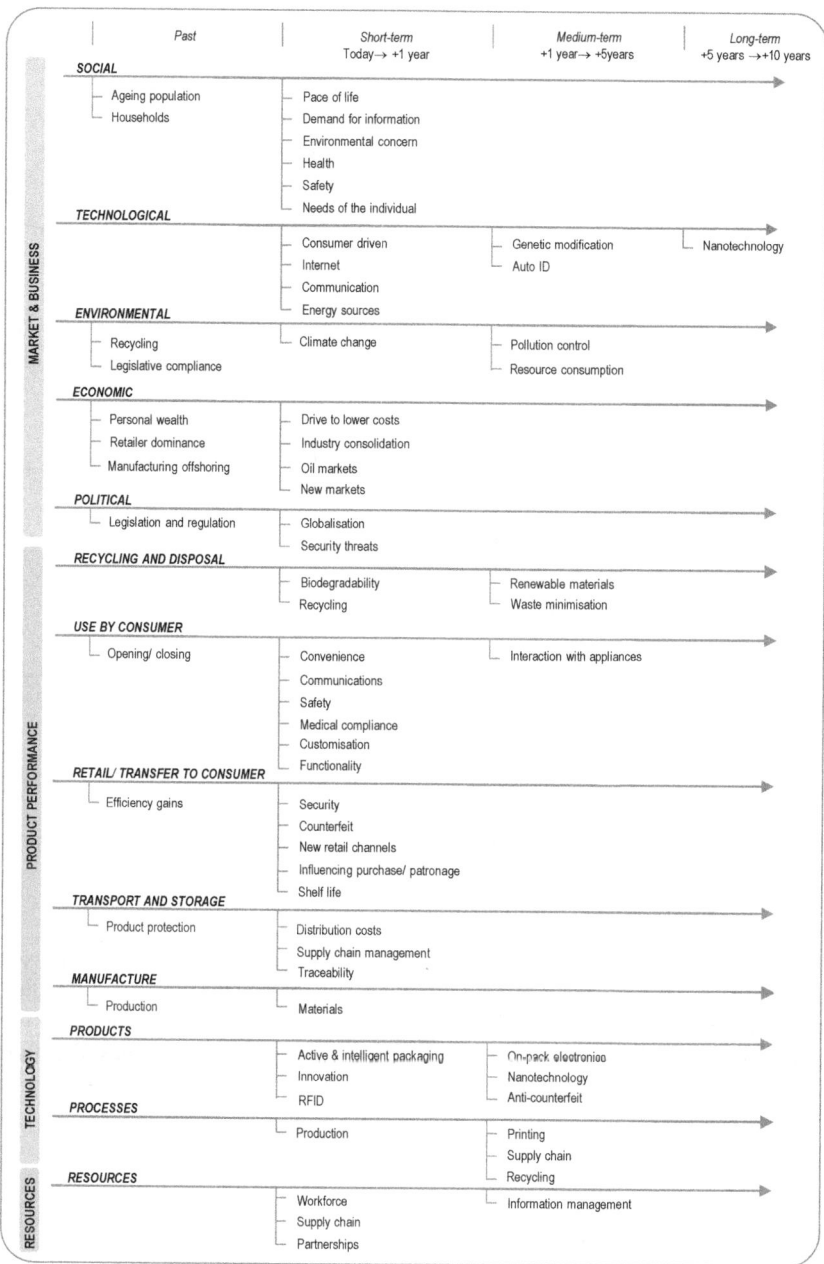

Fig. 6 The high-level depiction of the roadmap illustrating the abstracted codes in the layers of the roadmap architecture

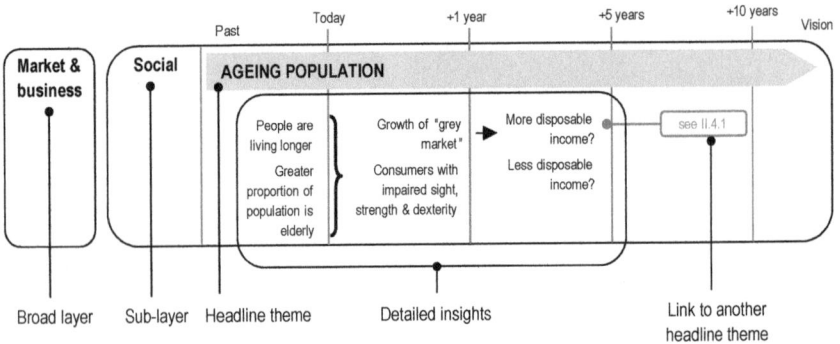

Fig. 7 An example of the information presented in the graphical representations of the detailed insights

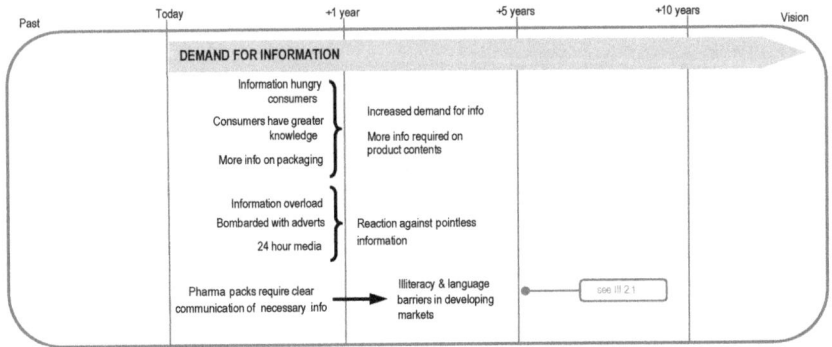

Fig. 8 Example of how the detailed insights are illustrated and described in the roadmap

3.3.2 Developing a Working Document

There is often confusion as to whether a roadmap is an actual graphical/ visual representation of some form of strategic plan, or whether it is a written document or report that contains graphical/ visual representations. The final roadmap produced by this process is a report, which surveys and describes the future competitive landscape of the packaging sector.

4 Conclusion

The International Roadmap for Consumer Packaging is an example of how different organisations and professional disciplines can collaborate to explore future market, product, technology and resource trends. In particular it is an example of how a standardised exploratory roadmapping process can be applied to capture and structure insights to develop future views of the competitive issues facing a diverse industrial area.

The benefits of collaboration in supply chains and industrial sectors are widely acknowledged and are increasingly important as business environments become ever more competitive and complex. This is evident in the packaging sector, which is made up of a diverse array of organisations that face an extensive range of competitive challenges and opportunities. Although the specific mix of these challenges and opportunities is unique to individual companies and markets, there are some common factors that will affect the sector as a whole.

In order to obtain a better understanding of these competitive factors, insights were sought from organisations across supply chains, materials sectors, different markets and geographical regions. The common factors to emerge were used to develop an extensive list of the key trends and drivers facing the sector over the next ten years.

The roadmap is intended to be a resource that provides useful information, structure and context for strategic planning and innovation processes in the packaging sector. It presents a complex landscape of trends and drivers through which companies and supply chains will be required to navigate. The path that companies and sector-level organisations choose to take through this landscape will depend on individual priorities. However, through identification of a range of factors that are common to a variety of markets and products, the roadmap represents an initial step in developing a sector-wide response to meeting some of these key challenges and opportunities.

References

Beeton, D.A.: International roadmap for consumer packaging. Institute for Manufacturing. University of Cambridge, Cambridge (2006)

Beeton, D.A.: Exploratory roadmapping for sector foresight. PhD thesis, University of Cambridge (2007)

Beeton, D.A., Phaal, R., Probert, D.R.: Exploratory roadmapping for foresight. International Journal of Technology Intelligence and Planning 4(4), 398–412 (2008)

Osborn, A.F.: Applied imagination: principles and procedures of creative problem-solving, Scribner, New York, USA (1963)

Phaal, R., Farrukh, C.J.P.: Technology planning survey – results, Institute for Manufacturing, University of Cambridge, project report (March 14, 2000)

Phaal, R., Farrukh, C.J.P., Probert, D.R.: T-Plan: The fast start to technology roadmapping – planning your route to success. Institute for Manufacturing, University of Cambridge, Cambridge (2001)

Strauss, A.L., Corbin, J.: Basics of qualitative research: Grounded theory procedures and techniques. Sage Publications, London (1990)

Authors

David Beeton is a strategist and engineer who has held a number of roles to shape the future of blue-chip companies, entire industries, regional economies and breakthrough technologies. David is currently a Director at Urban Foresight, where his focus is on the potential of rapidly emerging sectors such as electric vehicles, smart grids and intelligent transport systems to accelerate the development of smart cities. David gained a PhD from University of Cambridge where he pioneered techniques to create exploratory roadmaps for industry foresight. He has since worked with industry and policymakers around the world to develop roadmaps for many different applications. This invariably combines his twin passions of innovation and sustainable development, deploying creative solutions that will have a positive and enduring impact.

Robert Phaal is a Principal Research Associate in the Engineering Department of the University of Cambridge, based in the Centre for Technology Management, Institute for Manufacturing. He conducts research in the area of strategic technology management, with particular interests in technology roadmapping and evaluation, emergence of technology-based industry and the development of practical management tools. Rob has a mechanical engineering background, with a PhD in computational mechanics, with industrial experience in technical consulting, contract research and software development.

David Probert is a Reader in Technology Management and the Director of the Centre for Technology Management at the Engineering Department of the University of Cambridge. His current research interests are technology and innovation strategy, technology management processes, industrial sustainability and make or buy, technology acquisition and software sourcing. David pursued an industrial career with in the food, clothing and electronics sectors for 18 years before returning to Cambridge in 1991.

Part 4: Linking Technology Roadmapping to Other Instruments of Strategic Planning

Technology roadmapping is a powerful method by itself, but at the company level even more benefit may be gained by linking it to other planning methods. Interaction with corresponding methods in strategic planning, innovation planning, product development, systems development and project management are of particular interest.

In part 4 of the book some of these links will be presented:

- *Vinkemeier* shows how technology roadmapping may be used as a preliminary method for a performance measurement system based on a Balanced Scorecard. The starting point of this contribution is to translate the Roadmap into clear and measurable objectives and indicators, which clarify innovation activities at operative, tangible and manageable levels.
- *Doericht* focuses on a linkage between technology roadmapping and communication policy of a company. In his view it is helpful to provide relevant groups of interest with information about new technologies, products or services at the right time. The starting point of his contributions is an explanation of a special mix of methods that has been developed at Siemens AG over many years, based on the proven "Picture of the Future" approach. This method used by Siemens also helps their customers and partners to orientate their portfolios toward both global megatrends and current technological trends.
- *Lee* gives an overview of how technology roadmapping relates to patent analysis. She describes where patent analysis supports decision-making at various points of the roadmapping process. The use of patent analysis to support business planning, product, technology and R&D planning is described.

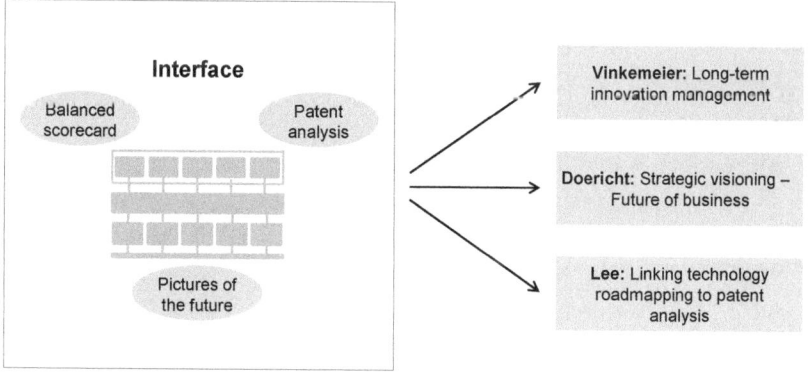

Fig. 1 Linking technology roadmapping to other planning methods

Long-Term Innovation Management – The Balanced Innovation Card in Interplay with Roadmapping

Rainer Vinkemeier

Being continuously innovative is vital for a company's success, which today is driven by globalization and fast innovation cycles. Even though all big companies operate their own innovation units, it is the effective management of innovations that separates the champions from the vast majority. By utilizing and combining well-established instruments of management and innovation, namely the Balanced Scorecard (BSC) and roadmapping, this paper presents a method for mastering said challenge. While roadmaps provide important information about distant future scenarios (markets, customers, products, technologies, etc.) and corresponding development paths (technology, products, skills) in terms of the company's future preparation, the innovation-adjusted BSC - now called Balanced Innovation Card (BIC) - translates the roadmap into clear and measurable objectives, indicators, measures, thus refining innovation activities on the operative, tangible and manageable level.

1 Introduction

Roadmaps may facilitate long- and mid-term orientation in technology, but they are far from being an instrument of operational innovation management. Innovation management requires much more - a linkage between roadmaps to assess those valuable though weak signals emitted by markets and technology on the one hand, and the management of operational resources for research and development on the other hand (Laube, 2007). An instrument which provides this is the Balanced Innovation Card as described here in interplay with technology roadmaps. A consequent connection of both instruments enables the establishment of a high consistency between strategic, long-term orientation and operative implementation in the field of innovation.

Rainer Vinkemeier
C21 Consulting GmbH, Kirchgasse 2,
65185 Wiesbaden,
Germany
e-mail: r.vinkemeier@c21-consulting.de

2 Implementing Technology Roadmaps: A Management Point of View

Let us commence with some warming-up questions.

1. How does a CEO convey to an employee in R&D - who frequently sees himself as a scientist rather than an executive organ within the company - that the corporate strategy, this abstract entity in whose creation he was not involved, is the model for everyday action?
2. How does the R&D guy in turn substantiate his original contribution to the company's long-term success and thus to the implementation of the aforesaid strategy?
3. How does he document the effect of his work on value-oriented financial indicators such as CFROI (Cash Flow Return on Investment) or EVA (Economic Value Added)?
4. How is it generally possible to induce groups as different as R&D, Marketing/Sales or the Finance Department to systematically and jointly exchange ideas on the levers of business and demonstrate cause-effect relations?

Not at all? Or at least not by means of a single instrument? False!

The Balanced Scorecard (BSC) - a management instrument established in Eurostoxx- and DAX-listed companies for well over a decade - is basically aimed at this kind of task, although not necessarily in R&D or the field of innovation on the whole. The BSC is rather used as a means of transporting the headquarters' strategy to business units.

However, the BSC is not an instrument designed for the long-term outlook that is required in the innovation field. Therefore, the BSC has to be transferred and adapted to this area with its very specific fundamental conditions and planning horizons very carefully. If so, it can be connected to a long-term business perspective which, depending on the industry covers a period ranging from five to 15 years. For the latter set of tasks mentioned above, technology roadmapping is available as an acknowledged instrument of strategic management (Möhrle and Isenmann, 2005). Roadmaps can enable a systematic and far-reaching look into the future of a business area. Hence, this instrument is mostly implemented prior to a BSC in terms of the time perspective and its utilization. Consequently there is a point of intersection between roadmaps and BSC regarding time and content.

The following relates to the crucial question of this topic: How can BSC and roadmaps be combined in an overall concept of long-term innovation management, which comprises and supports the complete chain of effects (from the early tracking of weak signals to the point of operative implementation in the R&D department)?

For a better access to this problem, it makes sense to recapitulate the basic concept and ideas of the BSC, including preliminary perspectives and their linkage to the financial perspective. The subsequent section contains an appraisal of the BSC's triumph in the business world. A description of its extensive use serves to point out differences in the interpretation, purpose and scope of the BSC.

Turning to the innovation focus, the applicability of the BSC in innovation units will be analyzed, leading to the development of the Business Innovation

Long-Term Innovation Management – The Balanced Innovation Card in Interplay 245

Card (BIC) which adjusts the company's BSC to innovation specific requirements in terms of objectives, indicators, target values and measures while still complying with the overall objectives and defined dimensions of the company's BSC. This BIC will be illustrated by means of a practical example.

Finally, the integral concept of BIC and roadmapping will be introduced. Here, the advantages and results of roadmapping are utilized to deduce the most effective dimensions of the BIC and to specify individual contents. This helps to ensure that the gap between (innovation) strategy and operative project work is bridged, and the "intangible" entity of innovation is broken down to a "local" operative level (Kaplan and Norton, 2004).

3 The Balanced Scorecard – Basic Facts

Now, how did the BSC first come into existence, where and what are its roots? Kaplan and Norton (1996), the "inventors" of the BSC detected considerable deficits in the strategic concepts of US-companies. A particular starting-point was the criticism that often the indicators used in companies exclusively focused on financial quantities. Apart from that it was established that

- visions and strategy do not prove to be implementable.
- the strategy is insufficiently linked with the set targets of departments, teams and employees.
- the strategy is not consistently connected with resources allocation.

Based on this fundamental criticism, a research project conducted by Kaplan and Norton with a dozen leading US-companies took place in the early 1990s to develop management systems that would meet the growing requirements of business reality.

Kaplan and Norton (1996) explain that - just as a pilot would never consider navigating a plane by means of only one instrument - a company manager who normally has to handle an extremely complex construct, should also have an adequately comprehensive range of instruments at his disposal.

On the face of it, the BSC may appear to be a structured collection of indicators. It is, however, far more if understood as a "management approach", and serves as a link between the development of a strategy and its implementation. The basic concept of BSC therefore completes the traditional set of financial indicators by the addition of so-called preliminary indicators - also known as performance drivers - combined with result indicators (Weber, Radtke and Schäfer, 2006).

Four common perspectives of the Balanced Scorecard (Figure 1).

- The financial perspective indicates whether the implementation of a strategy contributes to improving the result. Indicators in this respect are, among others, the return on equity or EVA. Here, financial indicators have a dual capacity. On the one hand, they define the financial expectation regarding the strategy. In addition, they represent the final targets of the other BSC perspectives which should basically be connected with the financial perspective through cause-effect relations.

- The customer perspective demonstrates the strategic objectives of the company regarding the customer and market segments for which, on the other hand, indicators, set targets and measures have to be specified.
- The internal process perspective first of all represents processes that are mainly relevant for the achievement of financial goals and customer-market targets. In this context, it is useful to take the entire value chain into account.
- The learning and growth perspective includes indicators which represent the infrastructure necessary to attain the objectives of the other perspectives. Here the necessity to invest in the future is clearly emphasized.

Result numbers and performance drivers have to be in balanced proportions to one another, as:

- Result numbers without performance drivers do not explicate how the results have been achieved. Furthermore, there is no early feedback on the successful implementation of a strategy.
- Performance drivers without result numbers may substantiate the attainment of short-term operative targets and improvements, but they will not show whether these improvements actually lead to an enumerable success and eventually to an improved financial performance.

All targets and indicators of the BSC have to be connected with targets of the financial/economic perspective. Each of the indicators considered in a BSC should be part of a cause-effect chain ending in a financial objective reflecting the strategy of the company. If the BSC is understood and used in this , it is much more than a recent compilation of isolated indicators; it rather has to specify how improvements in operative fields take effect on the financial performance, namely by means of higher sales figures, higher contribution margins or lower costs.

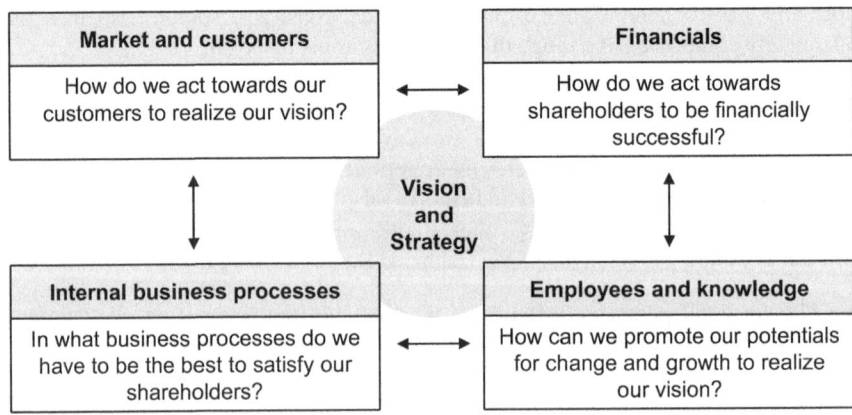

Fig. 1 Basic structure of the Balanced Scorecard

"The sequence of hypotheses on the cause and effect relations (among the individual indicators) ...has to be identified. Each criterion chosen for a BSC should be an element of such a chain of cause-effect relations, which makes clear to the company the importance of the company's strategy." (Kaplan and Norton, 2004).

Various obstacles may be overcome by use of the BSC:

- The development process of a BSC has to result in the clarification as well as to the consent as regards the strategic objectives.
- The BSC has to contribute to the consistent target orientation of the company's employees by means of three mechanisms: (i) communication and further education programmes, (ii) linking the BSC with objectives for teams and individual employees as well as (iii) linkage with incentive systems.
- In addition to human resources, financial and material resources also have to be brought into line with the company's strategy. The following steps mean to contribute to this: the formulation of intentionally high objectives, the identification of and the focussing on strategic initiatives as well as their linkage to annual budgeting processes.

Accordingly, the function of the BSC is to fully support the company's strategic management process and to serve as a scope of action for this process.

3.1 The Triumph of the Balanced Scorecard

There are hardly any executives - at least in big European companies - who are not familiar with the BSC. Many of them have already implemented it. Even though the intensity of dealing with this instrument may differ, it is astonishing how much has changed in the course of the past decade: The BSC has become both presentable and popular not only with controllers and diverse business developers, but also with line and functional managers.

The reason of its popularity lies in the intuitive reconstructability of the BSC's basic scheme (in fact, its proximity to common sense has often been pointed out) and the refreshing simplicity. Apart from that, a lot of managers say that previously they had felt unable to sufficiently relate individual pieces of improving information on markets and customers, (production) processes, personnel and innovation situation of their company.

Managers who are experienced users of the BSC consider the successful reconciliation of the gap between strategy and operative business to be its most significant benefit. In addition, they underline the productive dialogue involving all functions and hierarchic levels in the course of the joint formulation of the BSC (Kaplan, Norton, 2006).

3.2 Applicability of the Balanced Scorecard in Innovation Units

Particularly in areas like R&D and innovation - which are both crucial to the companies' long-term success - the BSC can display its beneficial characteristics.

However, in this context it is important to consider to what purpose and in which form the BSC is to be employed. Here, two basic types of BSC can be distinguished: firstly, the company or business unit BSC and secondly, the BSC of the functional units R&D/Innovation.

The contribution of the innovation unit to a company BSC comprises objectives, indicators and measures. This contribution shows that the innovation unit is an integral part of the strategy process of the entire company. The more the innovation unit is able to contribute, the more apparent its importance to the company as a whole becomes. Its relevance even grows if the BSC idea is successfully transferred to the innovation unit. This transfer is achieved by means of a specific R&D/Innovation- BSC. Based on the three elements of company BSC, innovation vision and innovation strategy, it has to be developed with utmost care. Objectives and measures already embodied in the overall BSC of the company or other business units can, in this context, become leading objectives within the BSC perspectives of the R&D unit. In this case it is important to group the other objectives of the individual innovation perspectives around them.

A BSC developed in this manner and interlinked with the company BSC is referred to as a Balanced Innovation Card (BIC) (Beeck, 2009). Meanwhile these functional BSCs have become established with respect to their definite relevance for management (Weber, Radtke and Schäfer, 2006). Examples of this type of BSC's development can correspondingly be found for innovation units.

In the case of a BSC for an innovation unit, all four BSC perspectives are bespoke to this unit's specific requirements, i.e. objectives, indicators, target values and, above all, measures must be chosen in a way that enables the support of the overall innovation objective derived from the company objective. A practical example is undoubtedly the best way to illustrate this mode of action. Therefore, a BIC tried in practice shall be presented in the following passage.

3.3 Example: Wireless Communication Technology

The innovation unit presented here by means of a BIC is a so-called development and application centre, part of a big German telecommunication company which is correspondingly active near end customer markets. The BIC comprises the four common perspectives: employees/knowledge, internal processes, market/customer and finance. Illustration 2 shows the variables which - depending on perspective - were identified as strategic objectives or drivers and transferred to the BIC. In this context, functions related to the innovation unit, such as Marketing/ Market Research and Logistics, were included in so far as they turned out to be critical with regard to the prospective success of innovations. Meanwhile, in the fifth release of this BIC, the focus has changed from corporate driven parameters to strictly innovation related parameters. For a delineation of this shift, release no.1 (2003/04) and no.5 (2009/10) are described in comparison in illustrations 2 and 3. During a routine process the BIC is developed in joint meetings, with participants from marketing, logistics and corporate finance. It takes three meetings in the course of three weeks' time to elaborate the BIC-outline, which is finalized by the management of the development and application centre as the leading unit.

Appropriate parameters are derived from each strategic objective. They are described by means of concrete measures or projects including the individuals in charge, so an immediate relation of all current and planned programmes is established.

The given example underlines how the BIC manages to display its coordinating and integrating effects with respect to the strategic management of innovation

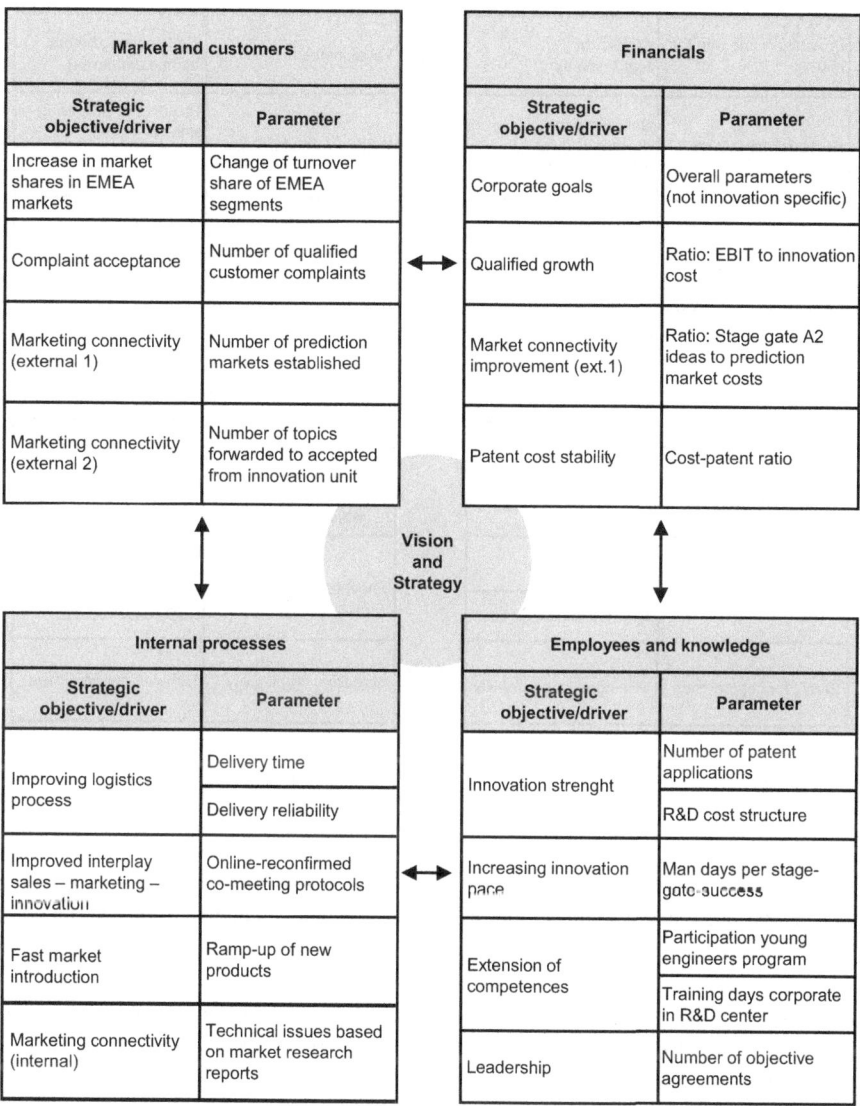

Fig. 2 Balanced innovation card in the field of wireless communication technology (Release 1.2)

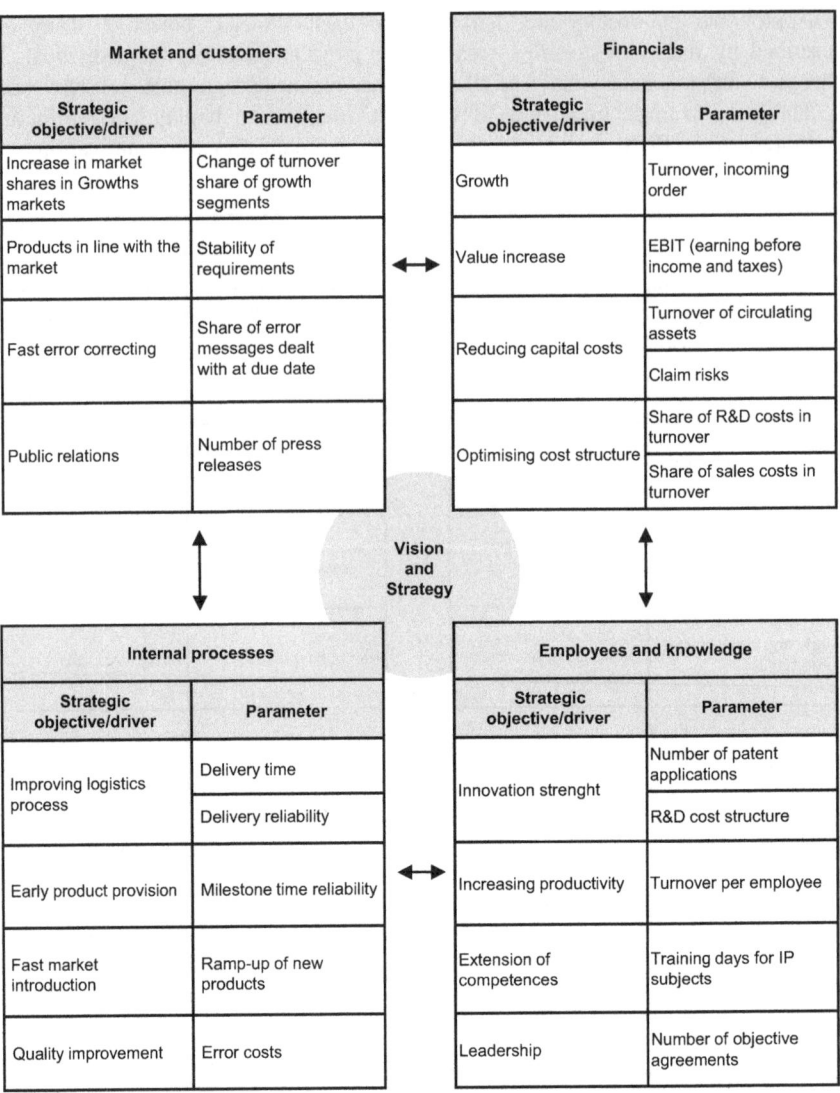

Fig. 3 Balanced innovation card in the field of wireless communication technology (Release 5.2)

units. Filling the gap between corporate strategy and operative business is also highly relevant to the aforesaid units. The BIC provides an instrument that is compatible with the instruments employed by superior units as well as by others on the same level. Since BSC s are developed in a hierarchic manner and can be "broken down", they dovetail with the companies´ current BSC architecture (Preißner, 2003).

3.4 Balanced Innovation Card and Roadmapping – An Integrative Concept

Initially, the question was raised whether the BIC could be a suitable element of an overall concept for the long-term management and controlling of innovations which represents the entire chain from the tracking of weak signals to an operative implementation. Depending on the respective branch, this chain covers a period ranging from three to 15 years, i.e. the lapse of time between the original "flash of genius" or fundamental invention and market introduction.

During the initial and usually rather extensive phase of this period, roadmapping is an appropriate instrument to support executives concerned with development in terms of a very structured, long-term perspective. On the basis of industry-specific scenarios and in accordance with most possible comprehensive expertises, roadmaps encompass a detailed record, assessment, selection and visualization of relevant market-related and technological developments in the form of so-called "development paths" (Weber, Kandel, Spitzner and Vinkemeier, 2005).

These development paths describe how the company or business unit ought to develop not only on the market or customer side but also with respect to the technology they represent, to live up to the future chances and risks that have been worked out in scenarios. Concrete statements can be derived from this as concerns:

- future market and customer structures and the products requisite for them,
- respectively, the necessary internal or external know-how,
- the corresponding investments and co-operations.

At the end of the roadmapping process, the resulting "innovation atlas" offers a systematic as well as far-ranging outlook into the future of one's business and thus the highest degree of transparency (Vinkemeier, 1999).

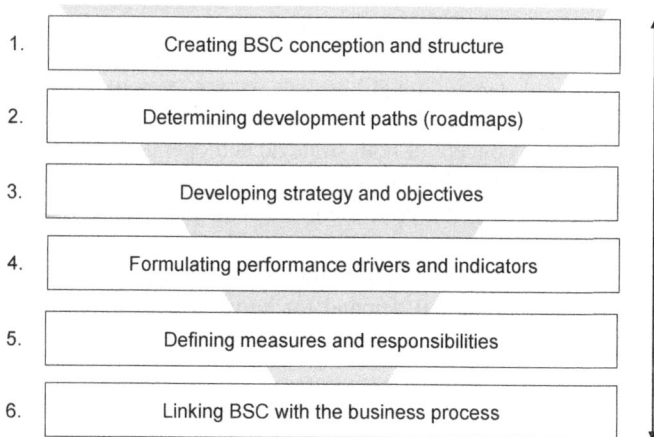

Fig. 4 Interplay between Balanced Innovation Card and roadmapping

What remains to be done now is the translation of this long-term transparency into concrete measures and actions. Although roadmaps provide information especially on upcoming planning periods with respect to products, technologies, know-how development etc., the gap between a strategic prognosis and operative project work is often difficult to bridge. However, if they are not seen as a purely academic exercise but intended to take effect in everyday practise - roadmaps must be integrated into the "local" operative work of innovation and R&D departments (Weber, Vinkemeier, 2007).

This is precisely where the BIC displays its efficacy. Namely by incorporating the strategy "cast" in roadmaps, concretizing it in its four perspectives and thereby "translating" the crucial elements of the roadmap into objectives, indicators, measures and individuals responsible for these measures (Figure 4). In order to understand how this happens, it is necessary to focus on the roadmap's crucial elements: the paths (products, technologies and/or skills). They are the core of each and every roadmap. Paths represent development steps and have the take in order to bridge the gap detected by internal and external experts (Vinkemeier and Franz, 2007). It would be impracticable and much too complex to focus on the whole innovation atlas as an outcome of the roadmapping process. Therefore, it makes sense to focus on selected paths with the highest relevance for economic or scientific reasons. As concerns these prioritized paths, it is important to highlight their relevance for, impact on and contribution to the market, internal process, knowledge and, of course, the financial perspective. Once they have been meticulously identified in the roadmapping process, they require careful management with the highest priority. Consequently, these paths are given a central position in the ongoing BIC-process. To be precise, they are "blended" into the dimensions of the existing BIC. They are integrated into the BIC process to enable their transfer from a primarily scientific or R&D-context and render them manageable in everyday business.

The following example means to illustrate the interplay between roadmapping and BIC

A manufacturer of aluminium components for the aerospace industry (i.e. OEM-supplier), is faced with the competitive pressure in this industry which has dramatically increased due to global mergers, as well as with shorter innovation cycles, and sees the necessity to re-orientate the company's innovation activities. The aim is a consistent concept of both strategic and operative management, i.e.:

- formulation of a conclusive innovation strategy on the basis of business unit scenarios, representing the concerted vision in the management team (time horizon of 12 years),
- adjustment of all R&D activities to long-term market necessities, especially: identification of the internal demand for know-how and creation of strategic partnerships for external know-how development (time horizon of eight to ten years),
- deduction of concrete innovation projects for the next two innovation cycles (time horizon of two or five years),
- co-ordinated management of these innovation projects, including the aspects of resources and project management, time to market, customer satisfaction/quality, retention and value orientation.

Long-Term Innovation Management – The Balanced Innovation Card in Interplay   253

In a first step, an innovation atlas is developed by means of a roadmapping process, which:

- contains all crucial steps for the company in terms of market and technology up to 2020,
- classifies the phases A: 2006 to 2010, B: 2010 to 2015, C: after 2015 and
- prioritizes the resulting development paths per business unit.

In phase A two development paths are categorized as being crucial for success: The development of alloys on a titanium basis up to market maturity, and the realisation of a consistent online management (CAS) of the main processes of casting, rolling, heat treatment and internal transport. Both development paths are to be supported by top management with special attention. Accordingly, the BIC is generated to provide a management instrument suited to this particular demand and purpose. The objectives of know-how development, command of processes, market visibility and the contribution to financial recovery are categorized as crucial for success with different weighting.

The BIC correspondingly comprises the four perspectives employee/knowledge, processes, customer/market and finances/value. The two highly prioritized development paths have a different effect on the four perspectives and correspondingly produce different partial objectives, parameters and measures. The development path "titanium" thus primarily aims at the BIC perspectives employee/knowledge and customer/market. In contrast, the development path "CAS" has a special effect on the process perspective and is shown here by means of indicators and measures. Therefore, these two paths are "blended" into corresponding BIC perspectives. In the perspectives of finance and value, the two strategic directions of impact merge again.

In this manner, roadmapping and the BIC complement each other to form a cohesive system of long-term innovation management. Impulses, weak signals and ideas for promising new products or technologies pass through a two-tier filter. In the end, signals that used to be weak are intensified, verified and assessed insofar as resources and responsible individuals can be allocated to them. Consequently, a maximum of transparency with regard to the roadmapping process leads to a maximum of attention and innovation intensity on the part of the company's top management where the BIC is concerned.

Now that the integral concept of BIC and roadmapping has been presented, several substantial conclusions remain to be drawn:

- The effective combination of acknowledged management and innovation instruments, i.e. the BSC and roadmapping, enable a bridging of the gap between long-term orientation, strategy and operative implementation in respect to innovation.
- Innovation units play a crucial part in business today. Hence, innovations units should continue to strengthen their role/identity inside the company through concrete action, e.g. demand and ensure their active participation /decision rights at a strategic top level by implementing a BIC or a roadmap.
- In order to ensure sustainable success, it is necessary to ascertain that generated BICs and Roadmaps find application and remain vital after implementation. Accordingly, their constant adjustment and enhancement should be embedded in the routine process.

References

Beeck, C.: Balanced Innovation Card: Instrument des strategischen Innovationsmanagements für mittelständische Automobilzulieferer. In: Ahsen, A.V. (ed.) Bewertung von Innovationen im Mittelstand, p. 125, 128, 134. Springer, Heidelberg (2009)

Kaplan, R.S., Norton, D.P.: The Balanced Scorecard: Translating Strategy into Action, p. 66, 144, 168. Mcgraw-Hill (1996)

Kaplan, R.S., Norton, D.P.: Strategy Maps: Converting Intangible Assets Into Tangible Outcomes, p. 89, 178. Mcgraw-Hill (2004)

Kaplan, R.S., Norton, D.P.: Alignment - Using the BSC to create Corporate Synergies, pp. 1–28. Harvard Business School Publishing (2006)

Laube, T.: Technology foresight and roadmapping. In: Fraunhofer IPA Conference and Workshop. Fraunhofer-Publica (March 2007)

Möhrle, M.G., Isenmann, R.: Grundlagen des Technologie-Roadmapping, Technologie-Roadmapping - Zukunftsstrategien für Technologieunternehmen, p. 5. Springer, Heidelberg (2005)

Preißner, A.: Balanced Scorecard anwenden, pp. 52–56. Hanser, München (2003)

Vinkemeier, R.: Roadmapping statt Glaskugel beim Blick in die Zukunft. Handelsblatt, No. 71, Daily Newspaper, p. b14 (1999)

Vinkemeier, R., Franz, M.V.: Controller und ihr Beitrag zum zukunftsorientierten Innovationsmanagement. In: Zeitschrift für Controlling und Management, Sonderheft, vol. 3, p. 42. Gabler, Wiesbaden (2007)

Weber, J.: Einführung in das Controlling. Schäffer-Poeschel, 73 (2002)

Weber, J., Kandel, O., Spitzner, J., Vinkemeier, R.: Unternehmenssteuerung mit Szenarien und Simulationen. Advanced Controlling 47, 21 (2005)

Weber, J., Radtke, B., Schäfer, U.: Erfahrungen mit der Balanced Scorecard Revisited. Advanced Controlling 50, 10, 13, 69 (2006)

Weber, J., Vinkemeier, R.: Controlling und Innovation. Advanced Controlling 56, 40 (2007)

Author

Rainer Vinkemeier is co-founder and managing partner of C21, a business consultancy specialized on future orientation and innovation management, located in Wiesbaden, Germany. His area of interest and professional focus since 1998 is to empower companies detecting and managing weak signals and early indicators announcing technological and/or market change. Roadmapping together with scenario technique, business simulation, and signal tracing are core areas of his work.

Strategic Visioning – Future of Business

Volkmar Doericht

Innovations are among the most important levers in corporate management. They safeguard competitive advantages, accelerate growth, and increase earning power. But what is the right path into the future? What technologies, what new business ideas should companies rely on so they can continue to meet their customers' needs in the future? Those questions are decisive for the competitiveness of companies. Strategic "invention of the future" as part of innovation management is needed now more than ever. Today, strategic planning of innovations means running a company's research and development so it is as precise and success-oriented as possible - in other words, running it effectively - and using financial means efficiently. All of this requires a clear understanding of technologies, customers' needs, and the markets of the future. This article explains a special mix of methods that has been developed at the Siemens AG over many years based on the proven "Picture of the Future" approach. Known as "Future of Business" this procedure is used by Siemens Corporate Technology to help its customers and partners within Siemens to orient their portfolios to both global megatrends and current technological trends. This method has been successfully applied to many different projects over the past three years, including "Future of Automotive 2020" (Siemens VDO), "Future of Commercial Transportation 2030" (Continental AG), and "Future of Airports 2030" (Siemens AG).

1 Introduction

The commercial success of the Siemens AG depends directly on its innovative strength. Corporate Technology is the strong innovation partner at Siemens for all sectors, divisions, business units and regions. The role of Corporate Technology is to develop technologies with which Siemens can distinguish itself as a trendsetter in the markets. Siemens uses the Picture of the Future approach to pursue three basic goals (Weyrich, 2002): First, to gain an overview of the technologies that will play a major role in the future. These are the ones that will generate future

---

Volkmar Doericht
Corporate Technology Innovation and Project Management,
Siemens AG, Otto-Hahn-Ring 6,
81739 Munich, Germany
e-mail: volkmar.doericht@siemens.com

market growth, then those that will have a multiple impact in many of the market sectors addressed by Siemens, and finally those that will lead to discontinuities, which means developmental breakthroughs. The second objective is to track down new business opportunities, and third, the intention is to communicate to people both inside and outside the company that Siemens is a visionary and innovative organization. This does not mean that ideas and visions are simply amassed; instead a systematic procedure is used which will lead within a reasonable period to market forecasts about the most important trends, the technologies underlying them, and ideas for new business opportunities. Examplarily, "The Future of Automotive 2020" project conducted in September 2005 presented trends in the automotive sector and was designed to serve as a source for business strategies. The most interesting new functionalities were autonomous driving (collision avoidance 2020), and the so called eCorner Module. This new drive system concept, which exclusively uses electrical and electronic systems, integrates not only the electric motor directly inside the wheels but also the steering, damping, and braking systems. This frees up space beneath the hood and eliminates the need for many attached parts in the steering column, brakes, and transmission. It thus opens up nearly limitless possibilities for automotive design. This completely new technologies have the potential to compete both technically and economically. In summer 2006, Siemens VDO engineers launched the first step on the strategic path with the electronic wedge brake that was expected to go into mass production at the end of 2010. Siemens VDO and competitor Continental AG supplied themselves up to the purchase of Siemens VDO by Continental in the year 2007 a race around the introduction of the series of this brake revolution, whereby the two companies pursued two different technical solutions. Above all, the Picture of the Future approach is intended to show how future objectives can be reached from the present. It is not all that important whether the forecasts of the future turn out to be right on target, because the world of technology is too dynamic for that, and the succession of developments is often turbulent. What is essential is the process that is described in detail below.

2 Theoretical Frameworks

It is the stated goal of Siemens to be the world's leading supplier of solutions for the great challenges that will result from global megatrends such as increasing urbanization or demographic change and which are associated with markets that will experience above-average growth over the long term. Siemens is orienting its operational business to those global megatrends. The new quality of orienting an organization to megatrends will be underpinned by the concept of strategic visioning. Strategic visioning (Embar, 1995) means actively shaping the future of society and business, the future of the organization, and the future of people and their own prospects in that organization. That is the origin of the Picture of the Future approach in the 1990's (Figure 1).

The analysis comprises two elements (Eberl, 2001): First, the current business is extrapolated and an attempt is made to derive forecasts for the future from it.

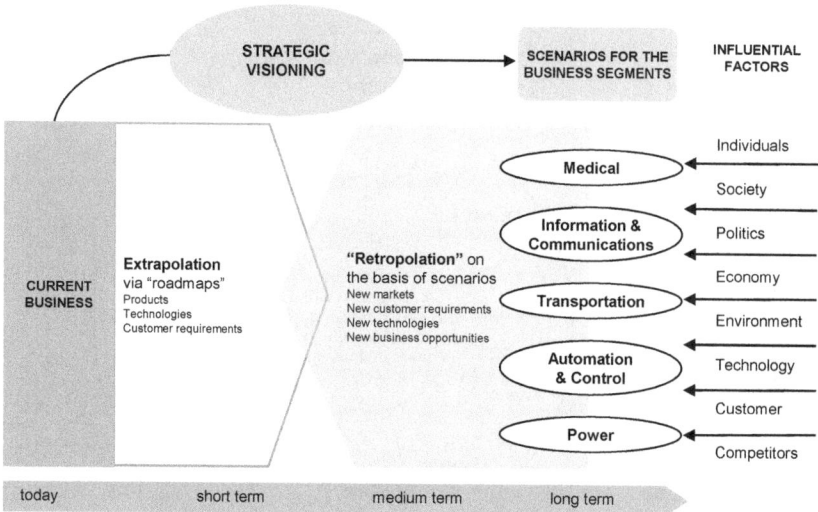

Fig. 1 "Picture of the Future" approach in 2001 (Siemens Corporate Technology)

In addition, regional scenarios and possible product concepts are developed from a series of socioeconomic variables. The requirements for future functionalities and technologies are in turn derived from them. The end result is the vision of a future product, a system, or a multi-system solution. This approach foresights what a market can develop into. The comparison of the two approaches generates which products and solutions are to be offered in the future. Specific medium- and long-term roadmaps for product development are prepared in this way, and gaps in the product and technology portfolio can be promptly identified and closed.

The early days of the "Picture of the Future" method date back to the mid-1990's. There was initially no uniform set of methodological modules. Hence, it was difficult to compare results and in some cases to understand the scenarios. A first step toward systematizing the method was to develop a granularity model. That model deals with the relativity of the concept of technology (see Figure 2).

A distinction is made between two different approaches to the development of products and solutions (Corsten, 1989): In "technology-driven product development," a new technical development comes first, followed by a search for new applications and users. In "demand-driven product development," technical development takes place in accordance with the needs of users. The model shown above includes both strategies. The left side is based on the needs of society or customers, which are satisfied by the core products of the operators and their suppliers. The core products are made using core technologies and protected by patents. On the right side, knowledge from fundamental or applied research is driven by technology-based development through design and prototyping to series production. Companies, like Siemens, in the manufacturing industry usually use

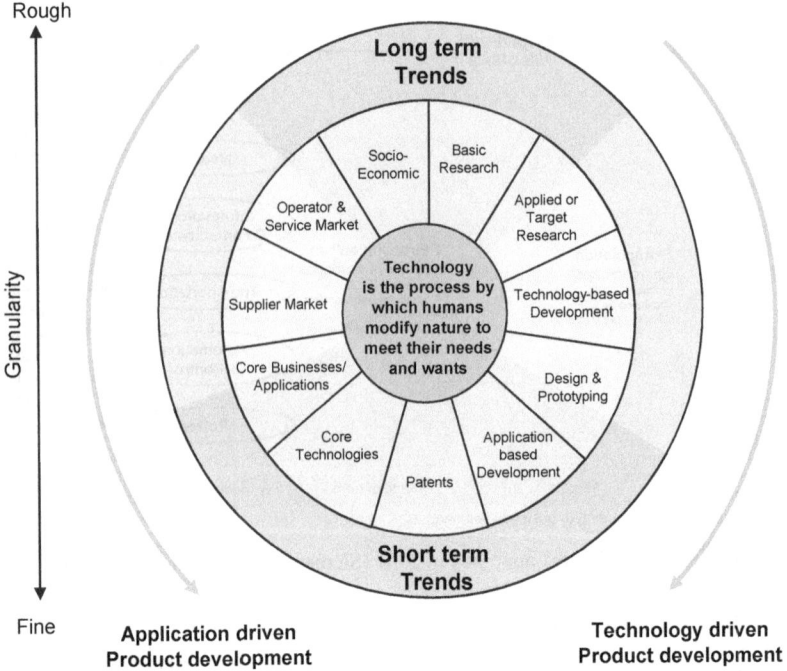

Fig. 2 Granularity model 2004

patents to protect their core technologies. Companies with "demand-driven product development" are more successful because the commercial success of the technical solution is based on societal acceptance (Schudy, 1999). Companies with "technology-driven product development" will be more successful as the degree of novelty of each product increases. In that case, the absence of any direct competitive pressure on the new technical solution during its early stages will have a particularly strong effect. The result of this is a latitude for pricing and the possibility of establishing an image as a benefit leader (Disselkamp, 2005). The Siemens AG is counting on a two-fold strategy for the long-term product and technology strategies of its sectors.

The primary role of the granularity model is to offer insights into the relativity of the period under consideration. The most important finding for the remarks below is that a rough granularity contains more stable long-term trends than a fine granularity. This means in turn that rough knowledge about the future is found particularly in the upper granularities. Taking advantage of this knowledge is the key to understand the future and leads to make the right decisions in the present. This characteristic also opens up new applications for existing concepts and methods. For example, it is still possible to use extrapolation in higher granularities even if it is no longer feasible to use it in the finer granularities. The following graphic results when this is applied to a timeline (Figure 3).

Strategic Visioning – Future of Business

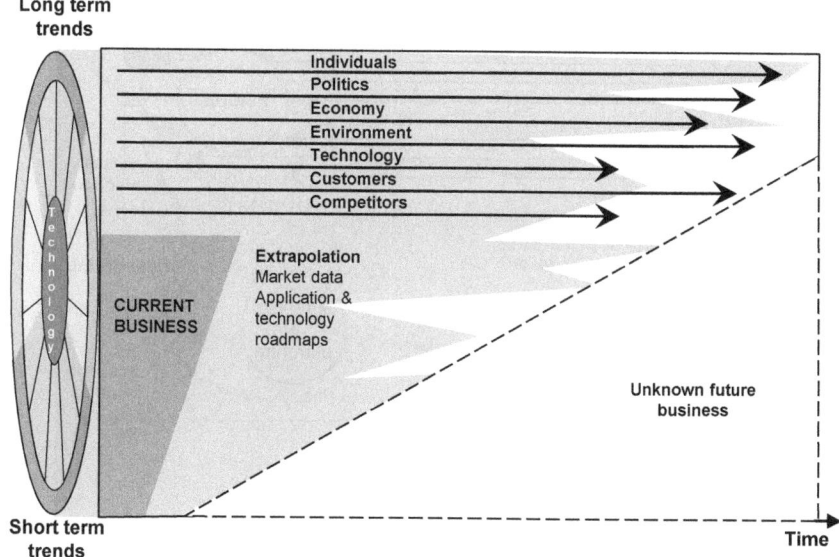

Fig. 3 Granularity model and exrapolation

Extrapolation, which is no longer appropriate for longer time periods at the level of current business, allows a breakthrough into the future at other levels of consideration, particularly to the level of socioeconomics. For example, multiple scenarios for the future can be formulated for a chosen longer time horizon, such as 20 years. An integrated approach takes into account as many long-term variables as possible, such as the development of societal, political, and industrial structures as well as sustainability, technological trends, and new customer needs. This knowledge about the future is thus based on stable long-term trends that are then used in several steps to develop scenarios that can be prioritized according to market attractiveness, risks, and opportunities. In that regard, a distinction is made among socioeconomic scenarios, industrial scenarios, product or application scenarios, and technology scenarios. The "Future of Business" method (Figure 4), in particular, shows an image of the future at very different levels of consideration. "Retropolation" (backcasting (Weaver et al., 2000)) is then used to identify tasks and issues which must be tackled today as the first definite step toward survival in the world of tomorrow. Consistent visions at the respective levels of consideration result from the combination of extrapolation and retropolation.

These visions help to quantify future markets, detect discontinuities, anticipate future customer requirements, and identify technologies with high growth potential and a widespread impact, as well as new business opportunities. By combining and consolidating the results of extrapolation and retropolation consistent strategies and plans can be developed. This knowledge is then visualized in product and technology roadmaps. Technology roadmapping is a

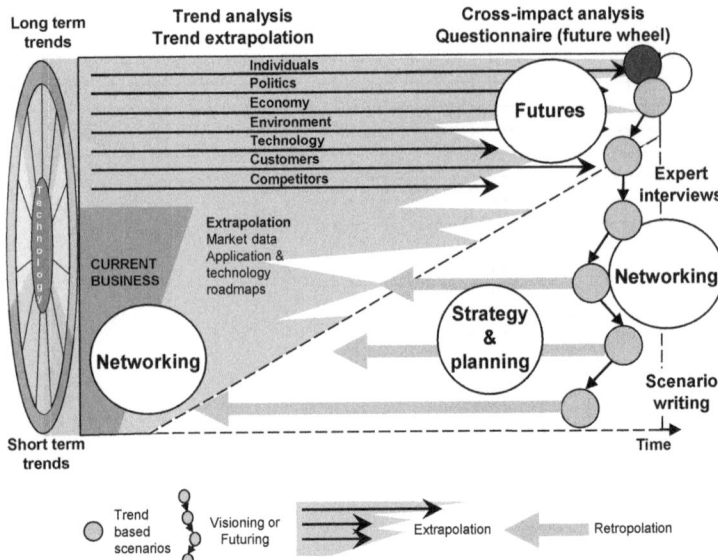

Fig. 4 "Future of Business" method 2005

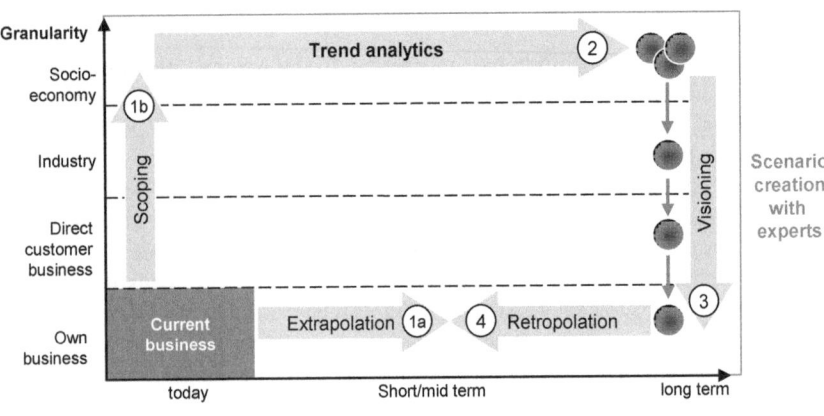

Fig. 5 The "Future of Business" (FoB) method 2008

powerful tool that enables to link technological capability to product and business plans so that strategy and technology development go hand-in-hand. A roadmap provides a graphical framework for exploring and communicating strategic plans. The "buy-in" and commitment of the key players to move forward the strategy and activities set out in a work plan. Technology roadmaps comprise a layered, time-based chart, linking market, product and technology information, enabling market opportunities and technology gaps to be identified. It has been developed by researchers at Cambridge University's Centre for Technology Management at the Institute for Manufacturing, and is used within the "Future of Business" (FoB) method (Figure 5) (Phaal et al., 2000, 2010).

3 Distinguishing from Other Concepts

Very different procedures can be used to plan for the future. For example, the Delphi method, brainstorming, and the scenario method are all used at Siemens. Extrapolation from current business and retropolation from future scenarios are also used. The particular advantage of the Siemens approach is that it is based on integrated procedures. When other concepts are used, they have weaknesses in either functionality or operationality. Any specific strategies for planning products and technologies which can be derived from their conclusions are therefore subject to limitations. Planning can be made more reliable only by using a structured, systematic method that, in addition to socioeconomic and industry-specific variables, also takes into account the full depth of all technological trends. For that reason, the Siemens AG has been strongly oriented to the concept of strategic visioning from the outset (Figure 1). David Sibbet (The Grove Consultants International) is considered the "father" of the strategic visioning concept (Karlöf, 2001). He has been working constantly since the 1960's to develop effective workshop methods in which language and images are combined. The focus is not just on that combination, but also on holding effective and, above all, results-oriented workshops that at once involve all participants and produce a workshop result that is supported by all participants. This approach integrates tools such as strategic planning practices, environmental analysis, SWOT analysis, and strategic and tactical goal-setting with the objective of developing and visualizing a necessary organizational change. Taken together, these elements lead to new quality in the thinking of groups all the way to a common understanding of future goals and personal roles and responsibilities. This basic philosophy is also behind the entire "Future of Business" method (Figure 6).

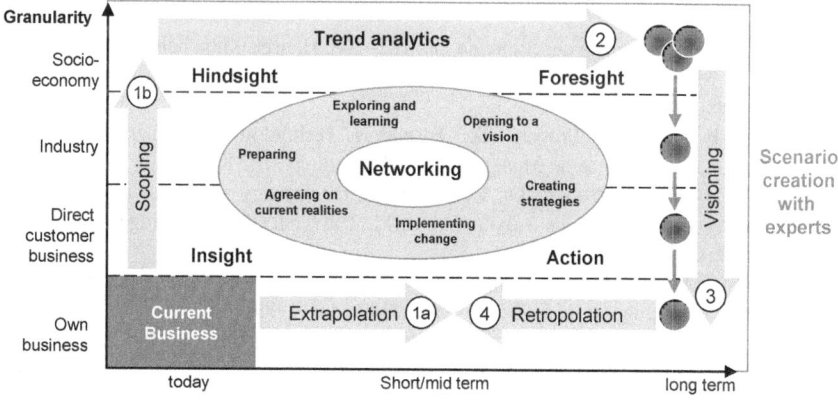

Fig. 6 Strategic visioning concept based on David Sibbet in the "Future of Business" method

4 Conclusion

In recent years Siemens Corporate Technology has worked with the Siemens business units to develop a set of highly effective instruments in order to develop and optimize strategies for innovations, systematically and sustainably. One result is known as the "Future of Business" method. It uses two opposing viewpoints that complement each other when they are combined: extrapolation from "today's world" and retropolation from the "world of tomorrow or the day after tomorrow." This method is not only carried out systematically, but is intended to be a process that can be used repeatedly. That is the only way that it can become an integral part of a culture of innovation throughout the company. "To transform the results into business success, however, you need much more: outstanding technological performance, a convincing portfolio of patents, efficient project management and - last but not least - world class teams. Ultimately they are the ones who will have to stand on the playing field of the market and successfully score goals." (Weyrich, 2002)

References

Corsten, H.: Die Gestaltung von Innovationsprozessen: Hindernisse und Erfolgsfaktoren im Organisations-, Finanz- und Informationsbereich, p. 7. Erich Schmidt Verlag, Berlin (1989)

Disselkamp, M.: Innovationsmanagement - Instrumente und Methoden zur Umsetzung im Unternehmen, p. 64. Gabler, Wiesbaden (2005)

Eberl, U.: Pictures of the Future - ein Verfahren, die Zukunft zu erfinden. Pictures of the Future, p. 5. Siemens AG (Fall 2001)

Embar, C.: What is Strategic Visioning?, Software Engineering Institute (1995), http://www.stsc.hill.af.mil/crosstalk/1995/11/Strategi.asp (accessed 2009)

Karlöf, B.: Die Renaissance der Strategie, pp. 102–108. Hanser, München (2001)

Naisbitt, J.: Megatrends. Ten New Directions Transforming Our Lives. Warner Books, New York (1982)

Phaal, R., Farrukh, C., Probert, D.: Fast-Start Technology Roadmapping. In: 9th International Conference on Management of Technology, IAMOT 2000 (2000)

Phaal, R., Farrukh, C., Probert, D.: Roadmapping for strategy and innovation - aligning technology and markets in a dynamic world. University of Cambridge, Institute for Manufacturing (2010)

Schudy, J.: Technikgestaltungsfähigkeit - Untersuchungen zu einer neuen Leitidee technischer Bildung, pp. 155–156. Waxman, Münster (1999)

van Someren, T.C.R.: Strategische Innovationen: So machen sie ihr Unternehmen einzigartig, p. 57. Gabler, Wiesbaden (2005)

Weaver, P., Jansen, L., van Grootveld, G.: Sustainable Technology Development, p. 97. Greenleaf Publishing, Sheffield (2000)

Weyrich, C.: Eine ganzheitliche Zukunftsplanung, Pictures of the Future, p. 4. Siemens AG (Spring 2002)

Author

Volkmar Doericht is senior consultant with Corporate Technology, Product & Service Innovation in Munich. He is one of the leading experts for Strategic Visioning and Strategic Foresighting within the Corporate Technology of the Siemens AG. In this role he is responsible for building the know-how on innovation strategy and the coordination of internal, national and international projects. He regularly lectures on science and industry conferences and is involved in many international programs and organizations.

Linking Technology Roadmapping to Patent Analysis

Sungjoo Lee

Patents, which are public documents organized in standard formats, are regarded as a valuable source of technical and commercial knowledge about innovative technical processes and activities. Putting roadmapping techniques together with patent analysis can increase the objectivity and reliability of a technology roadmap, while using patent analysis restricted to technological information together with roadmapping techniques can ensure that a more valuable breadth of strategic information is extracted from patents. The two techniques can complement each other in the strategic planning process. This chapter describes how to apply patent analysis for roadmapping so that the patent analysis supports decisionmaking at various points in the roadmapping process.

1 Introduction

A technology roadmap (TRM) can be a more powerful tool for strategic planning when used together with other management instruments. Recent efforts have focused on linking technology roadmapping to other planning tools such as scenario mapping (Lizaso and Reger, 2004), quality function deployment (Groenveld, 1997), TRIZ (Moehrle, 2004), and on trying to expand its applications to wider areas including knowledge management (Brown and O'Hare, 2001), new product development (Petrick and Echols, 2004) and service planning (Ahn et al., 2008). One such effort is to link technology roadmaps to patent analysis. Patent information has long been used for strategic technology management, so as to assist the technical decision-making both of inventors and of firms performing R&D, as well as adding value to economic policy-makers' decision-making. It has also been used to measure the strengths and weaknesses of competitors' technology and to plan technology development activities, and is therefore seen as a suitable methodology for analyzing both trends in technology and business opportunities based on technological capabilities.

Sungjoo Lee
Department of Industrial and Information Systems Engineering,
Ajou University, Suwon, Kyungkido,
Republic of Korea
e-mail: sungjoo@ajou.ac.kr

Table 1 The complimentary roles of roadmapping and patent analysis

	Roadmapping	Patent analysis
Conventional approach	Market-driven (Restricted use of technological assets)	Technology-focused (Limited perspectives on business)
	Expert-based (Lack of reliability and objectivity)	Statistics-based (Lack of strategic knowledge)
	Concept-based (Few roadmapping guidelines)	Analysis-focused (Few interpretation guidelines)
Combined approach	Technology planning with commercial perspectives	
	Strategic decision-makings based on statistical information	
	Guidelines for patent-based roadmapping provided	

Utilizing technology roadmapping and patent analysis together can allow them to play complementary roles. The objectivity and reliability of technology roadmapping, which is usually based on experts' opinions, can be increased greatly by integrating patent analysis into the process, while using patent analysis restricted to technological information alongside roadmapping techniques can ensure a more valuable breadth of strategic information is extracted from patents. Linking the two tools has several advantages. First, it enables us to systemize the process of technology-driven roadmapping as well as market-driven roadmapping using patent information, which is one of the most representatives of technology assets. Secondly, it helps develop guidelines for roadmapping based on patent analysis. Finally, it allows us to interpret patent analysis results in the context of roadmapping, which incorporates strategically important commercial perspectives. In fact, while conventional patent analysis is helpful for gaining technological information and identifying the present condition of technology assets, it does not attempt to integrate technological and commercial perspectives so as to identify promising new business opportunities, but this is possible by combining patent analysis with roadmapping. Table 1 shows how the two tools can be combined to increase the effectiveness of strategic planning.

A patent map, which is produced by gathering patent information related to a target technology field and then processing and analyzing it, produces a visualized expression of the total patent analysis results. The resulting map facilitates the easy and effective understanding of complex patent information, and can therefore be used to help make strategic management decisions. This chapter describes how to apply patent analysis for roadmapping.

2 Patent Analysis for Strategic Planning

Patents save an exclusive right to use, manufacture and marketing an invention. Additionally they provide the most comprehensive and most current sources of technical knowledge. No other source of information is so detailed and finely divided, so that patent information is very important for operational planning and decision-making processes.

2.1 Basics of Patent Analysis

Patents, which contain very specific information about technologies, have long been studied since they are regarded as offering valuable data for studies of technological innovation and trends. Among the various approaches to analyzing patents, three - patent index analysis, patent citation analysis, and data-mining analysis - are the methods most frequently used in developing patent maps.

In patent index analysis, various indexes are designed to accord with the purpose of the analysis, and then values from patent documents are used as a reference for technology planning or development. For example, Patel and Pavitt (1997) used an RTA (Revealed Technology Advantage) index to assess technology levels, while Ernst (1998) used RPA (Relative Patent Activity) and RPP (Relative Patent Position) indexes to estimate R&D levels, and RGR (Relative Growth Rate) and RDGR (Relative Development of Growth Rate) indexes to estimate the degree of technological attraction. Research has also been conducted on collaborative patenting activities, where the degree of technological collaboration was estimated using two indexes for Internal Collaboration (IC) and External Collaboration (EC) (Yamin and Otto 2004), while many other studies have used patent indexes to describe the current status of technologies.

Patent citation analysis is a bibliographic method that allows diverse information to be retrieved from analyzing the relationships between the citing of patents (i.e., where previous patents are cited in a specific patent application) and patent citations (where a particular patent is referred to in subsequent patent applications. or in other literatures. The frequency of this (latter) patent citation can be used as a proxy measure to estimate the degree of the subsequent technological effects of the invention detailed in the patent (Karki 1997). The more cited the patent, the more 'leading edge' and central to a particular technology the invention is likely to be. So patent citation analysis can enable us to explore competitive technological activities (Engelsman and van Rann 2004), and has been widely used to gain information about subsequent effects. The time gap between patents being granted and their subsequent citation in other patent applications gives a hint as to the technological cycle time involved. Patterns of knowledge flow can also be observed through patent citation analysis: citation information has been widely used to analyze linkages between technologies, degrees of technological influence and the impact of new technologies, as well as the structure of knowledge networks within or between industries or nations.

To increase the scope of analysis and the richness of information that can be uncovered from patents, recent studies have applied text-mining methods to the description sections of patent documents. Text-mining is designed to uncover and visualize useful patterns in textual data (Losiewicz et al., 2000), and has widely been used to retrieve information from intellectual property data. Applying text-mining to patent documents usually involves identifying keywords in the documents, which are then used in one of two ways. First, they can be used to develop a keyword vector, where the frequency of the keywords' use in the patent documentation is assigned to a corresponding vector field for each patent so as to distinguished them and measure the similarities of their contents (Yoon and Park, 2004). Secondly, co-word analysis, which measures the frequency of the co-occurrence of keywords in patent documents can produce a visual mapping of the relationships between them (Ding et al., 2001), giving a visual mapping where similar keywords are connected to each other. When applied to patents, this yields good results for technological planning, uncovering 'hidden' relationships between product and technology attributes.

2.2 Patent Analysis for Roadmapping

Patent analysis can be used at various points of roadmapping for business planning, product and technology planning, and R&D planning. The role of patent analysis in each planning process is briefly introduced here.

Patents can be analyzed to find new business opportunities based on its technological capabilities, suggesting a new approach of technology-driven roadmapping. Most existing roadmapping approaches tend to be constrained by market-oriented perspectives: such an approach regards technology roadmapping as the set of activities beginning with the perception of a market opportunity and ending with R&D requirements. Although customer needs and the competition situation are critical factors in a firm's success, it is unsatisfactory that, by contrast, technological breakthroughs are given such low significance that their value risks being overlooked. Since technology innovation can begin whole new business paradigms and uncover whole new markets, technology opportunity ought to be as thoroughly investigated as market opportunity. Central to the search for technology opportunity is technology capability analysis, highlighting firms' technological strengths and weaknesses, which can affect both the areas in which firms choose to do business, and how successful they will then be in such areas. For example, if a firm needs to diversify its business area, a promising option would be to enter a sector where its existing technological assets enjoy high superiority, thus helping to ensure the most efficient use of its technological assets and increasing the possibility of business success (Lee et al., 2009).

Patent analysis can be a decision-making tool for roadmapping, taking a computer-based roadmapping approach. Patent documents can provide valuable information with which not only to analyze technological trends, but also to predict product evolution patterns and relationships between technologies or products, which can contribute to increasing the effectiveness of the roadmapping process. If the description part of patent documents is analyzed, from which objective and quantitative knowledge on product and technology attributes can be extracted, it will facilitate more effective planning for products and technologies. Here, keywords extracted from patents are target objects of mapping, designing keyword-based roadmapping (Lee et al., 2008).

Once technology planning is completed, the next step is R&D planning, deciding how to acquire planned technologies and predicting any risk or opportunities of developing them. During the process, patent analysis can be carried out to set development targets and evaluating patent infringement risk (Lee et al., 2007) since patents provide information about the most-up-to-date technologies. It can also be used to explore possibilities of technology transfer when technologies are successfully acquired (Park et al., 2010), which is investigated by the relationships between technologies based on patent citation data.

3 Patent Analysis for Business Planning

For business planning patent analysis can be determined technological and strategic information on current and future competitors. This allows quick insights into the structure and competitors and in the technological strength of each competitor which could support the own evaluation of new business areas.

3.1 Concept

To use patent analysis for business planning based on technological capabilities, a technology-driven roadmapping process is proposed, which has four planning layers, as shown in Figure 1.

Unlike conventional roadmapping, it is designed to start from R&D planning, going through technology planning and product planning, and end with market planning, with the aim of identifying and then developing business opportunities based on technology assets. At the first 'R&D planning' - stage, R&D targets and schedules are determined. As part of this stage, it is essential to examine technology trends and competitors' activities, and the monitoring module is designed to discover relations between firms based on their technologies and identify which other firms have been doing similar research and which are leading the industry. After potential R&D targets have been selected, detailed development plans are elaborated at the second - 'technology planning' - stage. Possible new technologies that could result from R&D are discussed, including such issues as how to acquire those technologies and when they might be expected to be realized. The collaboration module shows relations between firms based on the knowledge flows in their patents, allowing a focal firm to consider its chance of realizing its desired technology by collaborating with others. Once technology planning has been completed, the next step is 'product planning', to identify new

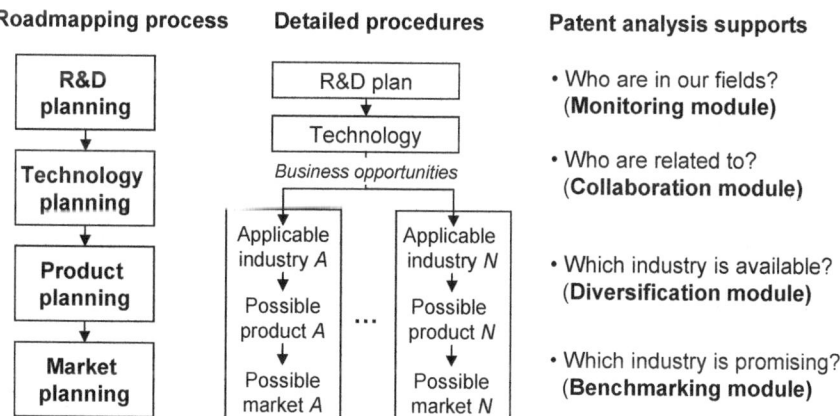

Fig. 1 Technology-driven roadmapping process

Table 2 Summary of patent analyses for technology-driven roadmapping

Modules	Questions	Patent maps
Monitoring	Who are in our fields?	Actor-similarity map
Collaboration	Who are related to?	Actor-relation map
Diversification	Which industry is available?	Technology-industry map
Benchmarking	Which industry has value?	Technology-affinity map

business opportunities based on the technologies that will become available, which is the core of the technology-driven roadmapping process. A single technology developed for a specific context may turn out to be applicable in various industries with minimum modifications, so the most important task at this stage is to use patent citation analysis to discover such industries - this is the aim of the diversification module. Product planning is completed when an idea of all the possible products that might result as a consequence of applying the technology to different industries has been generated. However, the diversification module results only reflect the technological aspects of possible product development avenues, but do not consider competitor activities or general market conditions and tends. Hence, the last stage - 'market planning,' - seeks to identify markets where other firms with similar technological assets are competing to understand the potential connections between products and markets. The benchmarking module is then employed to ascertain which other firms might be worth benchmarking. Through these four stages, the firm can finally decide the most promising market(s) where a particular technology can be best applied. Table 2 summarizes the patent analyzes relevant to each module - detailed descriptions follow in the next section.

3.2 Application Procedures

Four kinds of patent analysis are appropriate to support the different stages of technology-driven roadmapping, and the four associated patent maps are developed here.

3.2.1 Monitoring Module

For the first monitoring analysis an 'actor-similarity map' is developed (Figure 2(a)), which shows the relationships between the main actors in a specific business area based on their technological similarity by visualizing their patents' contents. This involves the following five steps:

- Step 1: Collecting all the patents for a specific technology;
- Step 2: Extracting keywords from these patent documents;
- Step 3: Constructing a keyword-vector for each patent;
- Step 4: Analyzing similarities between patents and then between firms;
- Step 5: Visualizing a network among firms.

Linking Technology Roadmapping to Patent Analysis

First, patent documents (or abstracts) are text-mined to extract keywords representing their contents using one of the various available software systems, which list the major keywords by their frequency or weight. After experts with domain knowledge have screened them to eliminate the meaningless ones, a keyword set is determined which will be used to measure the similarity of individual or groups of patent documents. Then allows a keyword vector is developed showing the frequency of each keyword in each patent. The next step is to analyze the similarity of patents from the Euclidian distance between keyword vectors. For example, the distance between [1, 2, 3] and [1, 3, 4] becomes $\sqrt{(1-1)^2 + (2-3)^2 + (3-4)^2}$. Euclidian distance is acceptable for measuring patent similarity when analyzing patent abstracts, since they tend to have similar length, but full documents are analyzed, some normalization of their lengths will be necessary. Finally, the similarity of the firms is analyzed based on this similarity of patents, producing a visual mapping of the relations between firms based on the assumption that the more similar two firms' patents are, the more similar their technologies will be.

3.2.2 Collaboration Module

The collaboration analysis yields an 'actor-relation map' (see Figure 2(b)), which is similar to the previous case, except that relationships are measured by technological knowledge flows rather than by technological similarity, and according to firms' general technologies in firms, rather than with regard to the specific technology being planned. Citation analysis uses a bibliographic method (commonly used to observe knowledge flow patterns) to uncover implications from the relations between citing of patents and patent citations to produce a map of actors' relationships. Taking the knowledge flow in patents as indicative of technological flows, many previous studies have used patent citations to analyze relationships between industries or nations based on R&D diffusion, and interfirm connections can be analyzed in the same way. Examined together with the results of the first map, this analysis can help a firm make strategic decisions about collaborative R&D, and the decisions can be applied in the early stage of roadmapping, especially for R&D and technology planning. The actor-relation map is developed by:

- Step 1: Collecting patents for an industry;
- Step 2: Analyzing patent citations by firms;
- Step 3: Visualizing a network among firms.

First, patents of interest are collected and classified by their applicant firms, and the total sum of patent citations for each firm is calculated to indicate the degree of knowledge outflow from that firm towards its competitors. In a similar way, the total sum of citing of patents is measured to indicate the degree of knowledge inflow from competitors. Finally, the values of citing and cited frequencies are represented on the similarity matrix where, for instance, c12 denotes that patents owned by Firm 1 cited patents in Firm 2 c12 times. This matrix is then used to develop a visualization of the patterns of knowledge flows between firms.

3.2.3 Diversification Module

With regard to diversification, a 'technological-industry map' is suggested (Figure 2(c)), to help identify other industry sectors where the technologies to be developed (or which already exist) might be applied. This process uses patent citation analysis to indicate the likelihood of technologies being applied in different industries by taking the knowledge flows in patents as indicating technological flows, and assuming that industries with more technological flows will offer greater possibilities for technology applications. If a particular industry is revealed as adopting a good deal of knowledge from a particular technology sector, it would appear to offer more opportunities to use that technology. Thus the map allows a firm to identify promising business areas where it might exploit its existing technological assets, or those it plans to develop. The technology-industry map is developed as follows:

- Step 1: Defining industries in terms of patents;
- Step 2: Collecting patents for a technology and identifying relevant industries;
- Step 3: Analyzing patent citations between those industries and the technology;
- Step 4: Identifying those industries that that are likely to be most affected by the technology.

The first step is to define industries and assign relevant patents to them: one common approach is to use patent classifications from the USPC (United States Patent Classification) to represent industries. The next step is to collect patents for analysis. All patents in the USPCs which had ever cited any patents originating from the firm can be retrieved to consider every single possibility of technology diffusion to industries. Then, knowledge outflows from the technologies in the firm to those industries represented by USPCs are examined by citation analysis to identify those industries highly affected by the technology.

3.2.4 Benchmarking Module

Finally, for benchmarking, a 'portfolio affinity map' of technologies is developed based on an affinity index (Figure 2(d)). If two firms have similar combinations of patents, it is likely that their technological assets may be similar and their affinity value is therefore high. The business areas of a firm's major competitors with high affinity index values are identified and analyzed for benchmarking. The patent index analysis is a typical example of the use of patent analysis, where various indexes are designed to accord with the purpose of the analysis, and the values then gained from patent documents used as reference points for technology planning or development. The set of patents in each technology category is used as a proxy measure for a firm's technology portfolio, and the affinity index measures the similarities or differences between firms in the set. An industry where many firms with similar sets of patents are already doing business offers good possibilities of further technology application, but is also likely to be an arena of keen competition; an industry where only few firms with similar technological assets are competing may be easy to penetrate, but the possibilities for successful technology application should be examined carefully. By analyzing those business areas where competitors with similar technological assets are engaged, a firm can gain valuable information for future product and market planning. Examined together with the technology-industry map, the portfolio affinity map can help identify promising business areas for exploration. The map is developed as follows:

- Step 1: Collecting patents for a technology and relevant industries;
- Step 2: Developing portfolio vector of technology assets regarding the firm's technologies and those of its competitors;
- Step 3: Analyzing the similarity between these portfolio vectors;
- Step 4: Selecting the firms with the highest affinity value;
- Step 5: Measuring the number of patents in other relevant industries for patents in each competitor

After patents for a particular technology have been collected, the technology is divided into its subtechnologies (again, defined via the patent classification system) and the relevant patents assigned to each. Then, a portfolio vector of technology assets is developed for each firm, taking the numbers of patents in each subtechnology as its elements. Using the portfolio vectors, the similarity between two firms is measured by Euclidean distance or Cosine similarity, yielding affinity index values which revealing the degree of firm's similarity in terms of their technology assets. Finally, those with the highest affinity value - where competitors with similar technology assets are already operating - are selected to explore as the other possible industry sectors a firm could consider.

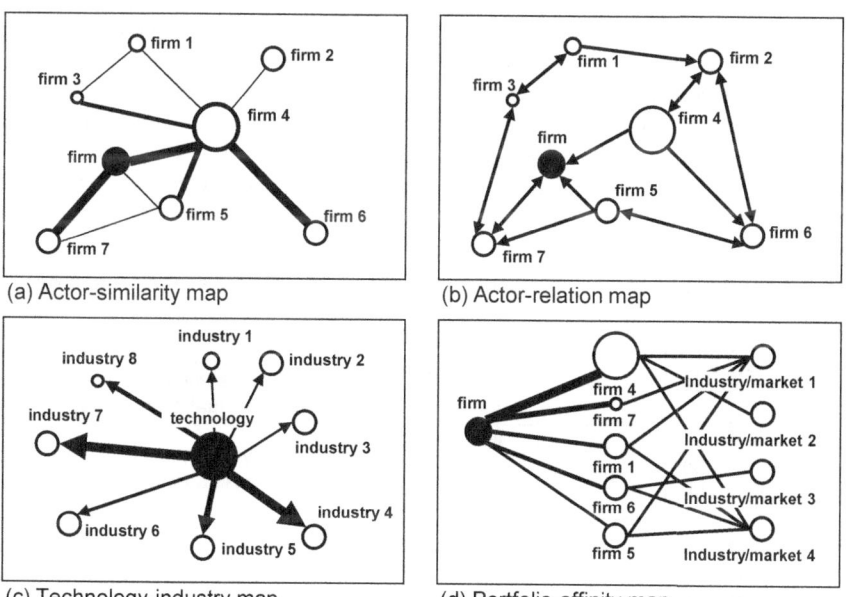

Fig. 2 Patent maps for technology-driven roadmapping

4 Patent Analysis for Product and Technology Planning

As for business planning the use of technological and strategic information on current and future competitors could also give an overview about the developed technologies of the competitors which helps to classify, align and plan the own technological developments.

4.1 Concept

While expert knowledge will often still play a decisive role - and may be the more desirable due to the strategic nature of technology roadmapping - objectivity is also valuable to technology roadmaps, and can be gained from quantitative information and its analysis. Taking a computer-based approach to roadmapping patent information can provide information that can ease the potential difficulties of the decision-making process, and provide an effective and systematic supporting tool in the product and technology planning stage. Information from experts about the evolutionary directions of products and technologies and relationships between and within product attributes and relevant technologies is especially vital in constructing a technology roadmap. The three patent analysis modules proposed here - the Evaluation, Investigation, and Dynamics modules - can be used to supplement or to check such expert opinion. Figure 3 illustrates how these three processes support roadmapping. Again, this analysis starts with the extraction of keywords from patent documents, and the elimination of meaningless elements. Those remaining are divided into two classes - product attributes and technology attributes - using product manuals and technology trees, and the keyword lists thus obtained are used to develop three different types of patent maps, each used to support decision-making in the roadmapping process.

The Evaluation module is used to provide information on major product development directions. When product planning is completed, we proceed to technology planning, where technology is incorporated into products. This process

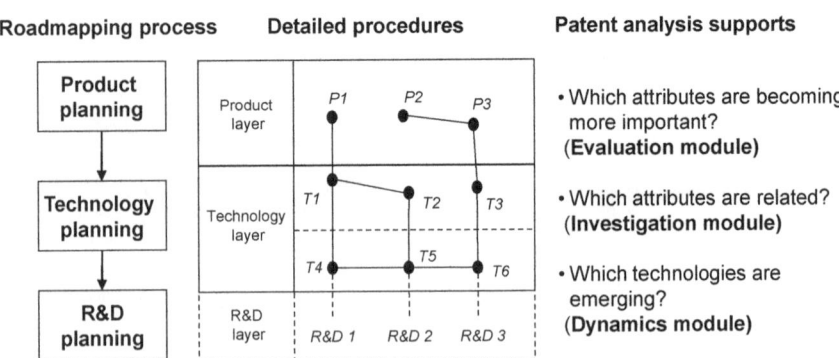

Fig. 3 Keyword-based roadmapping for product and technology planning

Table 3 Summary of patent analysis for keyword-based roadmapping

Modules	Questions	Patent maps
Evaluation	Which attributes are becoming important?	Keyword-portfolio map
Investigation	Which attributes are related?	Keyword-relationship map
Dynamics	Which attributes are emerging?	Keyword-trend map

is regarded as one of the toughest jobs in technology roadmapping, but the Investigation module can simplify the task by helping identify which technologies affect which product attributes, which of several alternatives technologies should be selected for application to a new product, and which new technology ideas can address particular product requirements. Finally, the Dynamics module is employed to illustrate technology trends, aiding the technology selection process. The patent analysis designed for each module is summarized in Table 3.

4.2 Application Procedures

Three kinds of patent analysis are designed for keyword-based roadmapping and three patent maps are developed to support decision-makings during the process.

4.2.1 Evaluation Module

For the first analysis - evaluation - a 'keyword-portfolio map' is developed (see Figure 4(a)), which measures the importance of product attributes and technology attributes. Developing this map involves:

- Step 1: Collecting patents and extracting keywords from their documentation.
- Step 2: Counting the keywords frequency in patent documents.
- Step 3: Developing a portfolio map for those keywords.

After keywords are extracted from patent documents, their absolute numbers and the rates of increase/decrease in their occurrence in patents are investigated, so that the keywords can be classified into four general types: core, emerging, established, and declining keywords. We assume that if keywords related to particular fields appear frequently in patents, that area is of significant interest among innovators in the relevant technology, and it is likely that R&D activities are in progress in these fields, with an increase in keyword frequency suggesting that they are becoming increasingly important. Core keywords appear frequently and show a relatively high rate of increase of occurrence. Emerging keywords occur moderately often but at an increasing rate. Established keywords are related to products or technologies already in the production stage and finally declining keywords, appearing with decreasingly frequency, are concerned with ebbing fields.

4.2.2 Investigation Module

In the investigation module, a 'keyword relationship map' is developed, which aims to effectively visualize meaningful relationships within product attributes, within technology characteristics, and between these two categories (see Figure 4(b)) through the following three steps:

- Step 1: Collecting patents and extracting keywords from them.
- Step 2: Analyzing the relationships between keywords using co-word analysis.
- Step 3: Developing a relationship map for those keywords.

In the analysis, the frequency of each keyword is calculated using the relevant patent data, and co-word analysis of these results generates a co-word matrix, where the value in the cell in ith row and jth column corresponds to the number of patents where the keywords i and j both words appear simultaneously. It is generally regarded that the simultaneous occurrence of two particular keywords indicates a relationship between them, and that such keyword relationships can be taken to illustrate corresponding relationships between product and/or technology attributes. The network analysis followed by co-word analysis produces results in the form of maps as in the figure, which are especially useful to identify and suggest which technologies need to be developed to improve the performance of certain product functions; which product attributes are highly related to each other and could therefore be considered simultaneously in new product design; and, when a particular technology needs improving, which others need to be considered at the same time. It also can be used to identify new technology ideas that involve combining several current technologies that affect particular product attributes.

4.2.3 Dynamics Module

This module uncovers the evolutionary patterns of products and technologies in patents and a 'keyword-trend map' is developed as follows (see Figure 4(c)):

- Step 1: Collecting patents and extracting keywords from them.
- Step 2: Developing a technology dictionary for the keywords.
- Step 3: Analyzing changes of keywords to produce a trend analysis.

After keywords are extracted from patents, experts prepare a technology dictionary linking product and technology attributes to relevant keywords. Based on this dictionary, changes of attributes in terms of keywords over time can be analyzed. In the technology layer, patents that include clear technological specifications can be especially useful. By illustrating the changing patterns of products and technologies in the past, this map can guide the direction of next-generation products.

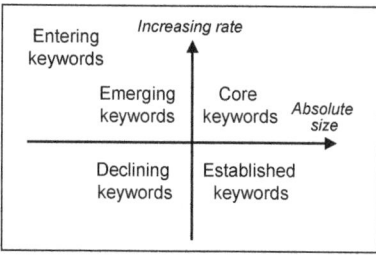

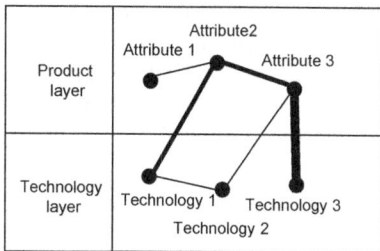

Fig. 4 Patent maps for keyword-based roadmapping

5 Patent Analysis for R&D Planning

Patent analysis is the key to many areas of an enterprise, from marketing, information and communication, distribution, development, human resources, the production to the legal department. Therefore, patent analysis has also an important role in R&D planning. For example, should an innovation be patented or not, how many patents should give, to plan what resources are needed, which licensing options are opportune, which countries should be protected or how the strategy profitable in the long term.

5.1 Concept

Patent analysis is especially useful for R&D planning. Specifically, it helps companies make decisions about how planned technologies can best be developed

Table 4 Summary of patent analysis for R&D planning

Modules	Questions	Patent maps
Transferability	Any ripple effect for technology?	Technology transferability map
Risk-assessment	Any risk in technology development?	IP risk map

and utilized. We suggest two kinds of analysis modules to deal with these issues: a Transferability module and a Risk-assessment module. The first aims to provide guidelines for transferring technology to developers so as to promote technological collaborations between various firms and facilitate the diffusion of R&D outputs, maximizing the external applicability of the developed technology, and thus ultimately increasing investment efficiency. The second module focuses on assessing development risk in terms of Intellectual Property (IP) infringement. The two modules are summarized in Table 4.

5.2 Application Procedures

In the following the procedures of the transferability module and risk-assessment module will be described.

5.2.1 Transferability Module

This module is designed to analyze the transferability of technology, so as to help facilitate the greatest use of R&D outputs in other industries, and proposes the use of patent information for analysis as follows:

- Step 1: Defining industry with respect to patents
- Step 2: Collecting patents related to a technology and relevant industries.
- Step 3: Analyzing patent citations between the industries and the technology.
- Step 4: Developing a portfolio map of relevant industries and identifying potential industries for technology transfer.

A 'technology transferability map' is designed and patent citation analysis, which reveals relationships between technology and industries, is used to address the transferability issue. Based on the analysis results, the potential industry aims to list several industries closely related to the technology from the point of both technology and industry. Industries to which the technological knowledge is given more are regarded as more significant from the technology point of view (knowledge outflow), while industries in which the technological knowledge holds an essential position are regarded as significant from the industry point of view (knowledge inflow). Generally, industries that are significant from both points of view are target industries, to which the technology is likely to be transferred. The analysis results are visualized through a portfolio map as shown in Figure 5(a).

5.2.2 Risk-Assessment Module

Investigating the existing technology is critical in planning R&D efforts not only to set development targets but also to identify development risk. For this purpose, an 'IP risk map' is suggested, which can be developed by the following four steps:

- Step 1: Collecting patents.
- Step 2: Measuring a technological importance (TI) index.
- Step 3: Measuring a patent importance (PI) index.
- Step 4: Developing a portfolio map of patents to avoid infringe

Patents of which management should be particularly aware will meet two criteria. The first is that the patent is technologically advanced and relates significantly to the firm's current development objectives. So, a patent is of value according to whether it fits in with what the firms wants to do next and is assessed by experts' opinion, producing a value for its TI value. The second is that the patent itself should be of high quality. The indexes most often used to measure patent quality include citation frequency, granted status, technological scope, international scope etc.: in combination, these amount to a PI index. Patents with high scores on both indexes should command management's attention (see Figure 5(b)).

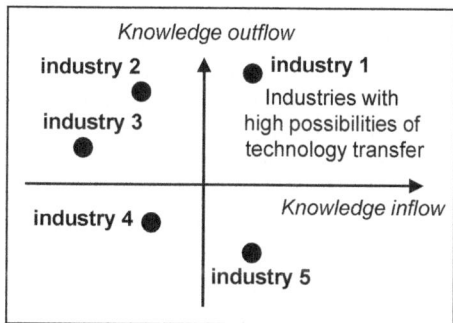

(a) Technology transferability map

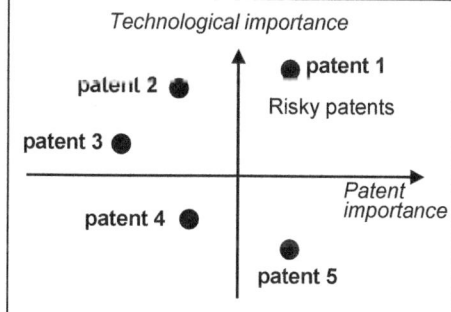

(b) IP risk map

Fig. 5 Patent maps for R&D planning during roadmapping

6 Conclusions

The effectiveness of technology roadmapping will only be increased when it is not used in isolation from other management tools, so designing how to integrate technology roadmapping into other tools for strategic planning is a prerequisite for effective roadmapping. As most roadmapping procedures depend heavily on the qualitative judgment of technical experts, the time of all the individuals involved in developing technology roadmaps represents a major contributor to roadmapping costs the higher the quality of results desired, the more time needs to be invested and the more these costs increase, risking total development costs being beyond firm's capacity. To increase roadmapping effectiveness, scattered information must be collected from various sources and transformed into appropriate form to support quick and accurate decision-making. Out of the various possible sources of such information, this article focuses on patents and suggests that linking roadmapping to patent analysis can greatly increase its effectiveness, with patent information supplying quantitative knowledge that can be a valuable objective data source to complement subjective expert opinion.

However, analyzing patents has typically involved considerable time and costs. Many commercial software systems have been suggested to counter this problem, and while their functionalities are generally limited to basic statistics, some offer surprisingly high levels of analysis. Wherever possible, we suggest the use of software to automate analyses, leaving only the final decision-making to be made by managers and experts. Nevertheless, while it may be possible to reduce experts' work by developing a computerized supporting system, their analysis and interpretation as to domain knowledge is indispensable, not only during keyword selection but also elsewhere in the research process. For example, user inputs are required to set cut-off values in developing the maps: where a user wants to see only core links between entities, higher values can be set, while lower values will reveal more complex and detailed pictures. Finally, while patents can provide valuable information, that information itself is subject to several limitations. Search results may not completely represent the whole picture: depending on the inventor's strategic purpose, some inventions may not be patented, and keyword searches may not identify all relevant patents, since firms are often less than transparent in their patent titles or abstracts to stop them being retrieved too easily.

For further work, more types of patent maps will need to be designed to reflect user needs at the various points of roadmapping. Novel IT techniques - such as data- and text-mining - have allowed us to extract more valuable knowledge from patent documents and to visualize the extracted knowledge more easily: it can be expected that further such advances will continue to make roadmapping both more effective, and more cost-effective.

References

Ahn, Y., Lee, S., Park, Y.: Development of an integrated product-service roadmap with QFD: A case study on mobile communications. International Journal of Service Industry Management 19(5), 621–638 (2008)

Brown, R., O'Hare, S.: The use of technology roadmapping as an enabler of knowledge management. Institution of Electrical Engineers 7, 1–6 (2001)

Ding, Y., Chowdhury, G.G., Foo, S.: Bibliometric cartography of information retrieval research by using co-word analysis. Information Processing and Management 37(6), 817–842 (2001)

Engelsman, E.C., van Raan, A.F.J.: A patent-based cartography of technology. Research Policy 23(1), 1–26 (1994)

Ernst, H.: Patent portfolios for strategic R&D planning. Journal of Engineering Technology Management 15(4), 279–308 (1998)

Groenveld, P.: Roadmapping integrates business and technology. Research Technology Management 40(5), 48–55 (1997)

Karki, M.M.S.: Patent citation analysis: A policy analysis tool. World Patent Information 19(4), 269–272 (1997)

Lee, S., Kang, S., Park, E., Park, Y.: Technology roadmapping for R&D planning: Case of parts and materials industry in Korea. Technovation 27, 433–445 (2007)

Lee, S., Lee, S., Seol, H., Park, Y.: Using patent information for designing new product and technology: Keyword-based technology roadmapping. R&D Management 38(2), 166–188 (2008)

Lee, S., Yoon, B., Lee, C., Park, J.: Business planning based on technological capabilities: Patent analysis for technology-driven roadmapping. Technological Forecasting & Social Change 76(6), 769–786 (2009)

Lizaso, F., Reger, G.: Linking roadmapping and scenarios as an approach for strategic technology planning. International Journal of Technology Intelligence and Planning 1(1), 68–86 (2004)

Losiewicz, P., Oard, D., Kostoff, R.: Textual data mining to support science and technology management. Journal of Intelligent Information Systems 15(2), 99–119 (2000)

McCarthy, R.C.: Linking technological change to business needs. Research Technology Management 46(2), 47–52 (2003)

Moehrle, M.G.: TRIZ-based technology-roadmapping. International Journal of Technology Intelligence and Planning 1(1), 87–99 (2004)

Park, Y., Lee, S., Lee, S.: Using patent information for R&D planning in multi-technology industries: The Korean aerospace industry case. The Journal of Technology Transfer (2010), doi:10.1007/s10961-010-9181-8

Patel, P., Pavitt, K.: The technological competencies of the world's largest firms: Complex and path-dependent, but not much variety. Research Policy 26(2), 141–156 (1997)

Petrick, I.J., Echols, A.E.: Technology roadmapping in review: A tool for making sustainable new product development decisions. Technological Forecasting and Social Change 71(1), 81 100 (2004)

Yamin, M., Otto, J.: Patterns of knowledge flows and MNE innovative performance. Journal of International Management 10(2), 239–258 (2004)

Yoon, B., Park, Y.: A text-mining-based patent network: Analytic tool for high-technology trend. The Journal of High Technology Management Research 15(1), 37–50 (2004)

Author

Sungjoo Lee is an Assistant Professor at Ajou University in Korea. She holds her BS and PhD in Industrial Engineering, both from Seoul National University in Korea. After spending six months as a Senior Researcher in the Ubiquitous Computing Innovation Centre in Korea, she moved to the UK to work as a postdoctoral research fellow for a year. Her area of work is technology and intellectual property management, and her current research focuses on technology roadmapping, patent engineering and open innovation.

Author Index

Abe, Hitoshi 173
Beeton, David A. 225
de Vries, Meike 211
Doericht, Volkmar 257
Farrokhzad, Babak 211
Farrukh, Clare 13, 91
Gerdsri, Nathasit 191
Geschka, Horst 123
Hahnenwald, Heiko 123
Isenmann, Ralf 1
Kanama, Daisuke 151
Kern, Claus 211
Kerr, Clive I.V. 67

Lee, Sungjoo 267
Moehrle, Martin G. 1, 137
Nimmo, Geoff 47
Orilski, Simon 107
Petrick, Irene J. 31
Phaal, Robert 1, 13, 67, 91, 225
Probert, David R. 13, 67, 91, 225
Schuh, Günther 107
Vinkemeier, Rainer 243
Wemhöner, Hedi 107

CPSIA information can be obtained at www.ICGtesting.com
Printed in the USA
LVOW01*1011060414

380506LV00010B/190/P

9 783642 339226